Lecture Notes in Networks and Systems

Volume 21

Series editor

Janusz Kacprzyk, Polish Academy of Sciences, Warsaw, Poland
e-mail: kacprzyk@ibspan.waw.pl

The series "Lecture Notes in Networks and Systems" publishes the latest developments in Networks and Systems—quickly, informally and with high quality. Original research reported in proceedings and post-proceedings represents the core of LNNS.

Volumes published in LNNS embrace all aspects and subfields of, as well as new challenges in, Networks and Systems.

The series contains proceedings and edited volumes in systems and networks, spanning the areas of Cyber-Physical Systems, Autonomous Systems, Sensor Networks, Control Systems, Energy Systems, Automotive Systems, Biological Systems, Vehicular Networking and Connected Vehicles, Aerospace Systems, Automation, Manufacturing, Smart Grids, Nonlinear Systems, Power Systems, Robotics, Social Systems, Economic Systems and other. Of particular value to both the contributors and the readership are the short publication timeframe and the world-wide distribution and exposure which enable both a wide and rapid dissemination of research output.

The series covers the theory, applications, and perspectives on the state of the art and future developments relevant to systems and networks, decision making, control, complex processes and related areas, as embedded in the fields of interdisciplinary and applied sciences, engineering, computer science, physics, economics, social, and life sciences, as well as the paradigms and methodologies behind them.

More information about this series at http://www.springer.com/series/15179

Elżbieta Macioszek · Grzegorz Sierpiński
Editors

Recent Advances in Traffic Engineering for Transport Networks and Systems

14th Scientific and Technical Conference "Transport Systems. Theory & Practice 2017" Selected Papers

Springer

Editors
Elżbieta Macioszek
Faculty of Transport
Silesian University of Technology
Katowice
Poland

Grzegorz Sierpiński
Faculty of Transport
Silesian University of Technology
Katowice
Poland

ISSN 2367-3370 ISSN 2367-3389 (electronic)
Lecture Notes in Networks and Systems
ISBN 978-3-319-64083-9 ISBN 978-3-319-64084-6 (eBook)
DOI 10.1007/978-3-319-64084-6

Library of Congress Control Number: 2017947033

Printed on acid-free paper

This Springer imprint is published by Springer Nature
The registered company is Springer International Publishing AG
The registered company address is: Gewerbestrasse 11, 6330 Cham, Switzerland

Preface

Contemporary technical and technological development as well as the ever-growing mobility provide favourable conditions for designing innovative transport-related solutions. Consequently, transport systems and networks require new decisions to be continuously made with regard to planning, organization and control of traffic. These decisions entail the necessity of seeking increasingly efficient solutions.

This publication, entitled *Recent Advances in Traffic Engineering for Transport Networks and Systems*, provides an excellent opportunity to become familiar with the latest trends and achievements in the field of contemporary transport systems as well as traffic engineering challenges and solutions. It has been divided into four parts:

- Part 1. Recent Advances in Traffic Engineering and Travel Models,
- Part 2. Safety Analysis in Advanced Transport Networks and Transport Systems,
- Part 3. Determinants of the Development of Transport Systems in European Union,
- Part 4. The Problems of Optimisation of Transport Networks.

The publication contains selected papers submitted to and presented at the 14th "Transport Systems. Theory and Practice" Scientific and Technical Conference organized by the Department of Transport Systems and Traffic Engineering at the Faculty of Transport of the Silesian University of Technology (Katowice, Poland). The topics addressed in the book include the current problems of transport systems, among other subjects discussed. With reference to numerous practical examples, various novel solutions applied in traffic engineering have been proposed in the publication. They are considered to exert significant influence on increasing the functional efficiency of transport systems and networks, and their priorities include well-being and health of people, traffic safety, sustainable development of transport systems and protection of natural environment.

We would like to use this occasion to express our gratitude to the authors for the papers they have submitted and their substantial contribution to the discourse on the multiple challenges facing transport systems and traffic engineering in the contemporary world as well as for rendering the results of their research and scientific work available. We would also like to thank the reviewers for their insightful remarks and suggestions which have ensured the high quality of the publication.

Readers interested in the latest achievements of traffic engineering and the overall body of problems addressed in this field of expertise may use this publication as an extensive collection of scientific research results, diverse insights and comments as well as new approaches and problem solutions. With the foregoing in mind, we are hoping that all readers will find this book valuable.

September 2017

Elżbieta Macioszek
Grzegorz Sierpiński

Organization

14th Scientific and Technical Conference "Transport Systems. Theory and Practice" (TSTP2017) is organized by the Department of Transport Systems and Traffic Engineering, Faculty of Transport, Silesian University of Technology (Poland).

Organizing Committee

Organizing Chair

Grzegorz Sierpiński — Silesian University of Technology, Poland

Members

Renata Żochowska
Grzegorz Karoń
Aleksander Sobota
Marcin Staniek
Ireneusz Celiński
Barbara Borówka
Kazimierz Dąbała
Marcin J. Kłos
Krzysztof Krawiec

The Conference Took Place Under the Honorary Patronage

Minister of Infrastructure and Construction
Marshal of the Silesian Voivodeship

Scientific Committee

Stanisław Krawiec (Chairman)	Silesian University of Technology, Poland
Tomasz Ambroziak	Warsaw University of Technology, Poland
Henryk Bałuch	The Railway Institute, Poland
Roman Bańczyk	Voivodeship Centre of Road Traffic in Katowice, Poland
Mehmet Dikmen	Baskent University, Turkey
Domokos Esztergár-Kiss	Budapest University of Technology and Economics, Hungary
József Gál	University of Szeged, Hungary
Andrzej S. Grzelakowski	Gdynia Maritime University, Poland
Mehmet Serdar Güzel	Ankara University, Turkey
Józef Hansel	AGH University of Science and Technology Cracow, Poland
Libor Ižvolt	University of Žilina, Slovakia
Marianna Jacyna	Warsaw University of Technology, Poland
Nan Kang	Tokyo University of Science, Japan
Jan Kempa	University of Technology and Life Sciences in Bydgoszcz, Poland
Michael Koniordos	Piraeus University of Applied Sciences, Greece
Bogusław Łazarz	Silesian University of Technology, Poland
Zbigniew Łukasik	Kazimierz Pulaski University of Technology and Humanities in Radom, Poland
Michal Maciejewski	Technical University Berlin, Germany
Elżbieta Macioszek	Silesian University of Technology, Poland
Ján Mandula	Technical University of Košice, Slovakia
Sylwester Markusik	Silesian University of Technology, Poland
Antonio Masegosa	IKERBASQUE Research Fellow at University of Deusto Bilbao, Spain
Agnieszka Merkisz-Guranowska	Poznań University of Technology, Poland
Maria Michałowska	University of Economics in Katowice, Poland
Leszek Mindur	International School of Logistic and Transport in Wrocław, Poland
Maciej Mindur	International School of Logistic and Transport in Wrocław, Poland
Goran Mladenović	University of Belgrade, Serbia
Kai Nagel	Technical University Berlin, Germany
Piotr Niedzielski	University of Szczecin, Poland
Piotr Olszewski	Warsaw University of Technology, Poland
Enrique Onieva	Deusto Institute of Technology University of Deusto Bilbao, Spain

Asier Perallos	Deusto Institute of Technology University of Deusto Bilbao, Spain
Hrvoje Pilko	University of Zagreb, Croatia
Antonio Pratelli	University of Pisa, Italy
Dariusz Pyza	Warsaw University of Technology, Poland
Cesar Queiroz	World Bank Consultant (Former World Bank Highways Adviser), Washington, DC, USA
Andrzej Rudnicki	Cracow University of Technology, Poland
František Schlosser	University of Žilina, Slovakia
Jacek Skorupski	Warsaw University of Technology, Poland
Aleksander Sładkowski	Silesian University of Technology, Poland
Wiesław Starowicz	Cracow University of Technology, Poland
Jacek Stumpf	Department of Documentation GDDKiA Section Katowice, Poland
Andrzej Szarata	Cracow University of Technology, Poland
Tomasz Szczuraszek	University of Technology and Life Sciences in Bydgoszcz, Poland
Antoni Szydło	Wrocław University of Technology, Poland
Grzegorz Ślaski	Poznań University of Technology, Poland
Paweł Śniady	Wrocław University of Environmental and Life Sciences, Poland
Andrew P. Tarko	Purdue University West Lafayette, USA
Frane Urem	Polytechnic of Šibenik, Croatia
Mariusz Wasiak	Warsaw University of Technology, Poland
Adam Weintrit	Gdynia Maritime University, Poland
Andrzej Więckowski	AGH University of Science and Technology Cracow, Poland
Katarzyna Węgrzyn-Wolska	Engineering School of Digital Science Villejuif, France
Adam Wolski	Polish Naval Academy, Gdynia, Poland
Olgierd Wyszomirski	University of Gdańsk, Poland
Elżbieta Załoga	University of Szczecin, Poland
Stanisława Zamkowska	Kazimierz Pulaski University of Technology and Humanities in Radom, Poland
Jacek Żak	Poznań University of Technology, Poland
Jolanta Żak	Warsaw University of Technology, Poland

Referees

Marek Bauer
Przemysław Borkowski
Piotr Czech
Domokos Esztergár-Kiss
Michal Fabian
Robert Grega
Mehmet Serdar Güzel
Katarzyna Hebel
Peter Kaššay
Jozef Kuĺka

Contents

Recent Advances in Traffic Engineering and Travel Models

Vehicle Speed Impact on the Design of Efficient Urban Single-Lane Roundabouts

Hrvoje Pilko(✉) and Šime Šarić

Faculty of Transport and Traffic Sciences, University of Zagreb, Zagreb, Croatia
hpilko@fpz.hr, sime.saric@yahoo.com

Abstract. Implementing effective roundabouts in urban areas is a highly demanding task and requires optimizing traffic (operational) efficiency (TE) and traffic safety (TS) while considering geometric factors, traffic characteristics and local constraints. To capture a relationship between roundabout design elements, traffic demand flow and vehicle path speed through the roundabout, we studied American and Australian method on four urban single-lane roundabouts located in the Croatian capital of Zagreb. Comparison between design speed and measured vehicle speed show bigger deviations than previous studies. Preliminary validation of the used methods enables better understanding of its application for Croatian conditions. These preliminary results suggest that the used methods can capture the correlation between geometry, traffic flow demand and vehicle speed of urban single-lane roundabouts but with needed calibration for local conditions.

Keywords: Urban single-lane roundabout · Geometry design · Traffic (operational) efficiency · Traffic safety · Vehicle movement trajectory · Vehicle speed

1 Introduction

Implementing effective roundabouts in urban areas is a highly demanding task and requires optimizing traffic (operational) efficiency (TE) and traffic safety (TS) while considering geometric factors, traffic characteristics and local constraints. Studies of roundabouts in various countries, particularly of single-lane roundabouts in urban areas, have shown that proper design can significantly improve TE [1, 2], as well as TS parameters [3–5].

Vehicle movement trajectory speed is another parameter that has influence on TE and TS [3, 6–8]. To capture a relationship between roundabout design elements, traffic demand flow and vehicle path speed through the roundabout, we study existing methodology. Eastern and South-eastern European countries have outdated roundabout design guidelines [9–13]. Although these guidelines do provide partial recommendations, it fails to define rules when setting design speeds for roundabouts. In such cases designers rely on their own experience, examples from good practice, and foreign guidelines [3, 14].

Croatia has approximately more than 200 roundabouts (>60% lie in urban areas), and many of them deviate substantially from international standards for roundabout

E. Macioszek and G. Sierpiński (eds.), *Recent Advances in Traffic Engineering for Transport Networks and Systems*, Lecture Notes in Networks and Systems 21,
DOI 10.1007/978-3-319-64084-6_1

planning, design and modeling, which compromises their TE and TS [15]. The country has no tradition of systematically monitoring TE, TS, vehicle path speed and other key performance indicators of roundabouts, though the government has called for the building and reconstruction of roundabouts as part of its National Traffic Safety Plan 2011–2020 [15]. The most recent national guidelines stipulate where urban single-lane roundabouts should be built, what geometry they should have, and how capacity should be calculated [9]. However, the guidelines do not indicate what models should be used for analyzing vehicle path speeds nor common TE parameters. These characteristics and previous studies [14] make the country attractive for studying impact of vehicle design speed on urban single-lane roundabouts.

The focus of the research is to study the impact of calculated versus measured vehicle design speed when designing urban single-lane roundabouts. This is done by using the same methodology like in [14]. The American and Australian methods [16, 17] are validated using field data from four urban unsignalized single-lane roundabouts located in the Croatian capital of Zagreb. Nevertheless, the influence of design elements on the vehicle path speed, i.e. on the TE and TS level, will not be studied here.

The remainder of the paper is organized as follows. Section 2 describes the used methodology and study area data. Section 3 presents validated model results, and Sect. 4 discusses the results and implications for future work.

2 Methodology

2.1 Design Speed and Vehicle Path Radii

The study concept and methodology of calculating vehicle design speed for vehicle path radii trajectories is the same as presented in [14]. The roundabout design speed is the maximum allowable speed for which the total driving safety is guaranteed in free traffic flow at the roundabout, under optimum conditions and with proper maintenance of the roundabout driving area [16]. Design speed is dependent upon the roundabout geometry and calculated according to driving path radii:

$$V = \sqrt{127R(e + f_s)} \tag{1}$$

where:

V - design speed [km/h],

R - vehicle path radius [m],

e - pavement slope [m/m],

f_s - coefficient of friction between a vehicle's tires and the pavement [16].

The stability and safety of vehicle passing through the roundabout is defined by the adhesion between the tire and the pavement. The coefficient of friction (f_s) is calculated by means of the coefficient of friction for light vehicles (f_{sLV}) and heavy vehicles (f_{sHV}):

$$f_sLV = 0,30 - 0,00084 \cdot \sqrt{M_VLV} \tag{2}$$

$$f_sHV = 0,30 - 0,00084 \cdot \sqrt{M_V HV} \tag{3}$$

$$f_s = (1 - P_{HV}) \cdot f_sLV + (P_{HV} \cdot f_sHV) \tag{4}$$

where:

f_sLV - coefficient of friction for light vehicles [-],
M_vLV - average weight of light vehicles [kg],
f_sHV - coefficient of friction for heavy vehicles [-],
M_vHV - average weight of heavy vehicles [kg],
f_s - coefficient of friction for vehicles [-],
PHV - percentage of heavy vehicles [17].

Recommended maximum design speed values for vehicles entering the mini roundabout is 25–30 km/h, and for small single-lane roundabouts is 30–35 km/h [16]. In the design process where there is no traffic flow demand and are no marked traffic lanes, the vehicle path is characterized by three movement radii: entry, circulatory roadway, and exit radius. It is assumed that the vehicle width is 2.0 meters, and that a minimum distance of 0.5 m should be maintained from the center of the roadway or concrete curb and painted edge of the splitter island. The imaginary vehicle path line is 1.5 meters away from the concrete curb and 1.0 m away from the painted line of the splitter island [16]. The fastest vehicle path for negotiating the roundabout is a series of reverse paths (the right-side path is followed by the left-side path, and the right-side path). In cases when there is no central island, the operating path will be a straight line. The methodology for defining the fastest vehicle path speed does not provide really expected vehicle operating speeds, but rather a theoretically possible speed of vehicle entry into the roundabout that is needed during the roundabout design [16]. Real vehicle operating speeds may greatly differ for various reasons, including different axle loads and vehicle characteristics, individual driver capabilities, and tolerance to gravity forces [16].

The speed consistency contributes to greater level of TS by reducing the speed difference between the conflicting streams of vehicles and optimizing entry capacity. That is why five critical radii must be checked for each approach: R_1 - the entry path radius; R_2 - the circulating path radius; R_3 - the exit path radius; R_4 - the left-turn path radius; R_5 - the right-turn path radius (Fig. 1). These vehicle path radii are not the same as the curb radii [16]. During design, R_1 should be smaller than R_2, and R_2 should be smaller than R_3, for the fastest vehicle path. This ensures that speeds will be reduced to their lowest level at the roundabout entry and will thereby reduce the likelihood of loss-of-control crashes. It is acceptable for R_1 to be greater than R_2 (maximum difference < 20 km/h). The R_4 radius must be evaluated to ensure that the maximum speed difference between the entering and circulating traffic is no more than 20 km/h. The design speed for radius R_5 should therefore be the maximum design speed for the entire roundabout and should not exceed the design speed of R_4 by 20 (km/h, as R_4 has a conflicting point with R_2 [16].

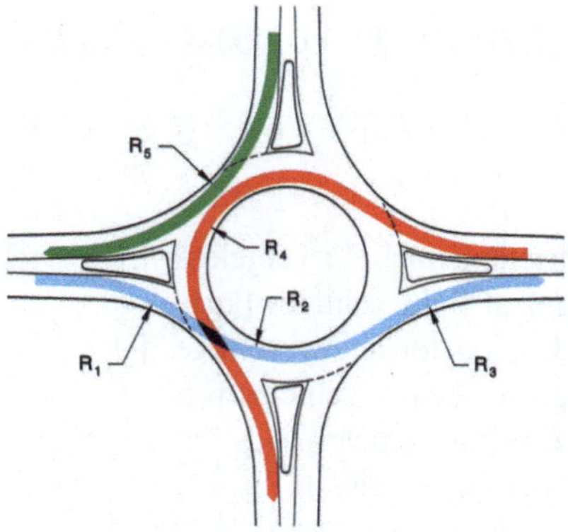

Fig. 1. Vehicle path radii [16]

2.2 Study Area and Data

Based on previous studies [14, 18] we apply the same methodology and repeat it on the same four urban single-lane roundabouts situated in the City of Zagreb. Their main design elements are given in Table 1. The roundabout Sveti Duh – Kuniščak has three approaches and all other roundabouts four approaches. Roundabout classification, their function of TE and TS in the city traffic network and layout (i.e. main geometry elements, speed limits of 40 km/h, number of traffic lanes, locations of speed bumps) of the studied roundabouts has not been changed since the last study. On some locations, new residential and office buildings have been build. Under normal traffic flow movements, the roundabout entry speeds (V_1), circulating speeds (V_2) and exit speeds (V_3) were calculated for the radii (R_1, R_2 and R_3). Path speeds for radii (R_4 and R_5), i.e. for the right-turn movements (V_4) and left-turn movements (V_5) through the roundabout, were not measured to assure the same study approach as in [14]. Traffic flow demand was measured by "pen and pencil" on 29th September 2016 (Thursday) in the morning peak hour traffic. At same day and hour, the entry, circulating and exit speeds were collected in 15-min intervals using the GPS (Global Positioning System) device installed in a passenger car (same vehicle and driver as in [14]). Favorable weather conditions enabled good visibility and dry pavement at all roundabouts and their approaches.

Table 1. Design elements at studied roundabouts [14].

Roundabout name	ICD [m]	CID [m]	w [m]	e [m]
Sveti Duh - Kuniščak	20.0	6.0	7.0	3.5/3.6
Petrova - Jordanovac	25.0	12.0	6.5	3.5/4.5
Voćarska - Bijenička	22.0	13.0	4.5	4.0
Radnička road – Petruševec 1	40.0	28.0	6.0	3.0/3.5

Legend: ICD - inscribed circle diameter; CID - central island diameter; circulatory roadway width (w), approach entry width (e).

3 Results

3.1 Results for 2016 Measurement

The sample size consists of again 50 measurements; average vehicle path radii and average path speeds at roundabouts. Design geometry data (i.e. pavement slope) were measured *in situ* by laser range finder (BOSCH PLR 50C) and designed with the help of Autodesk AutoCAD software. Data on pavement slope, percentage of heavy vehicles, and the corresponding coefficients of friction are calculated according to the latest traffic demand flow (Sect. 2.2) and presented in Table 2. Equations (1–4) were used to calculate design speed for negotiating a roundabout, while differences between the measured and calculated design speeds are presented and analyzed in Figs. 2, 3, 4 and 5. In Table 2, light vehicles are all vehicles belonging to categories L, M, M1, M2, and N, N1 and O1, O2, while heavy vehicles are vehicles belonging to categories M3, N2 and N3, and O3 and O4 according to [19].

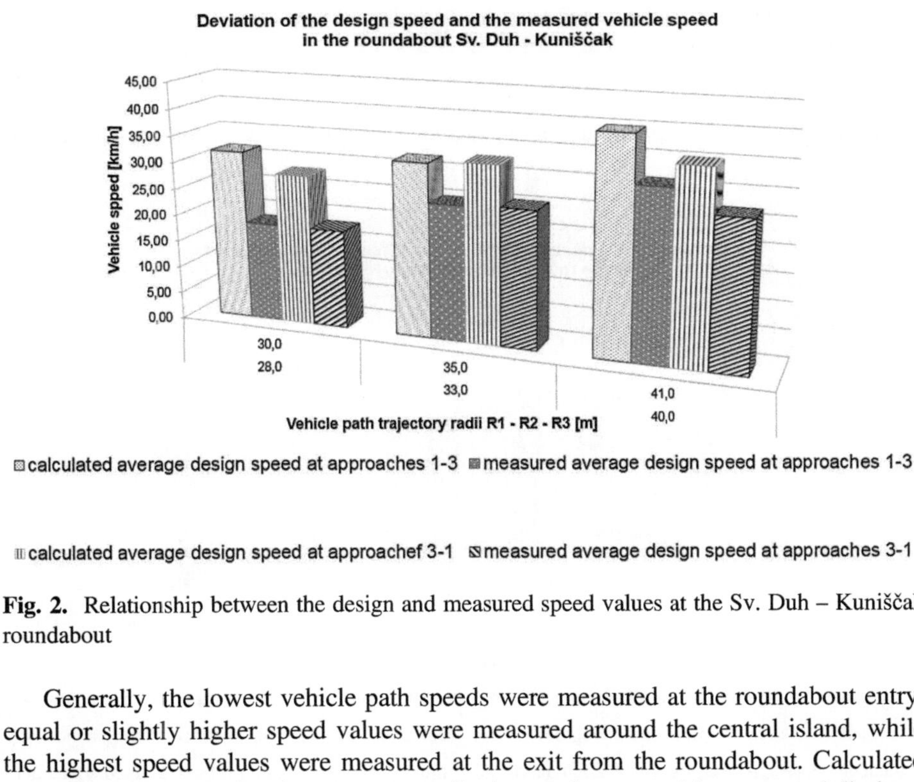

Fig. 2. Relationship between the design and measured speed values at the Sv. Duh – Kuniščak roundabout

Generally, the lowest vehicle path speeds were measured at the roundabout entry, equal or slightly higher speed values were measured around the central island, while the highest speed values were measured at the exit from the roundabout. Calculated average design speed values are generally lower than roundabouts speed limit of 40 km/h, and lower or slightly higher than the maximum recommended speed of 35 km/h, as shown in Table 2. All measured average speeds are lower than the recommended maximum speed of 35 km/h, and are also lower than the speed limit. Deviations between calculated average design speeds and measured average speeds and are presented in Table 2, and discussed as follows.

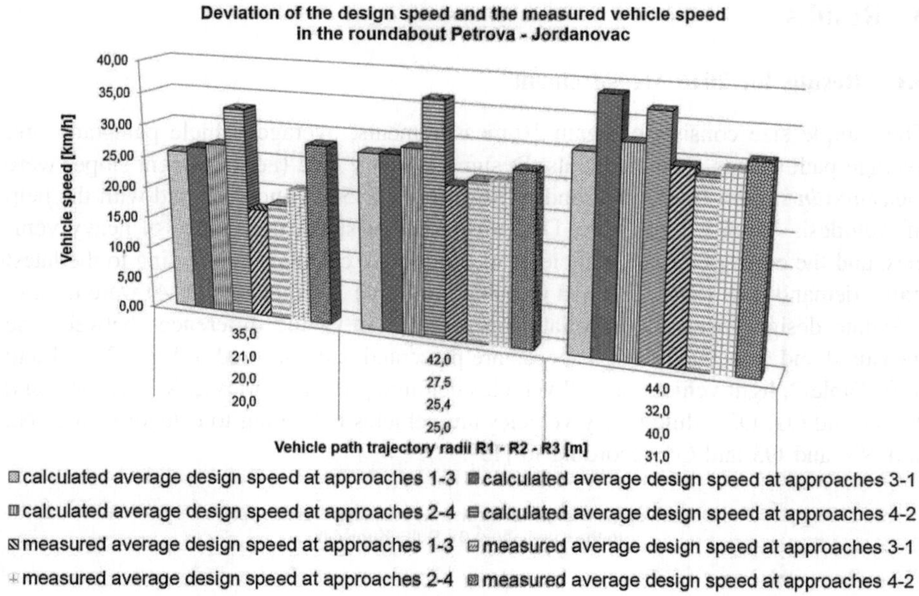

Fig. 3. Relationship between the design and measured speed values at the Petrova – Jordanovac roundabout

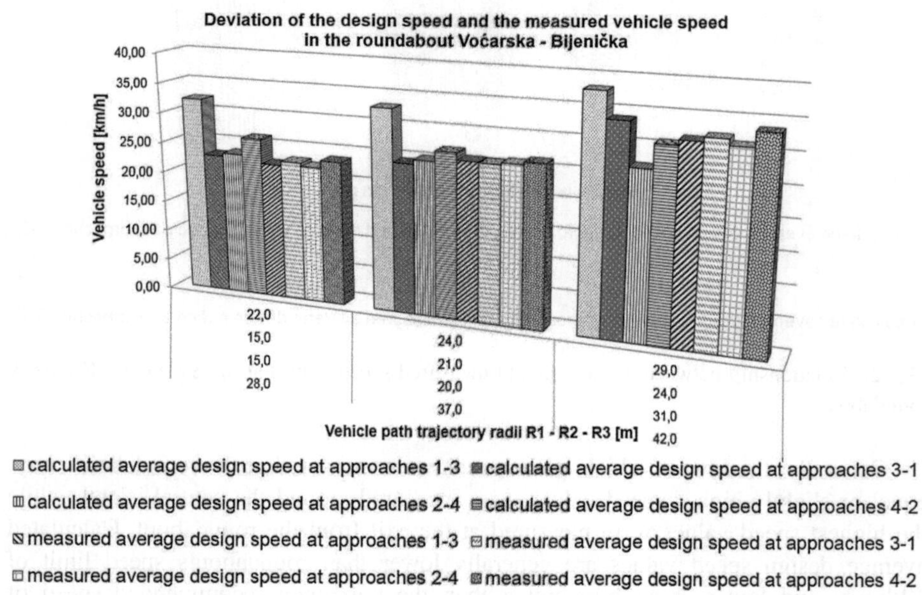

Fig. 4. Relationship between the design and measured speed values at the Voćarska - Bijenička roundabout

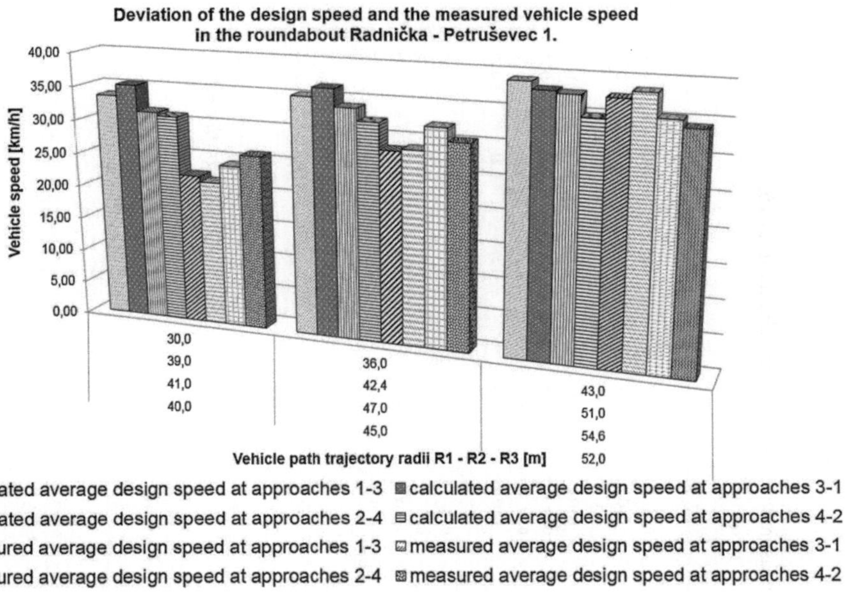

Fig. 5. Relationship between the design and measured speed values at the Radnička – Petruševec 1 roundabout

An average vehicle speed measured at the Sv. Duh – Kuniščak roundabout on the path from approach 1 (Sv. Duh – South) to approach 3 (Sv. Duh – North) was lower by 42.15% that the calculated average design speed. An average vehicle speed measured at the Petrova - Jordanovac roundabout on the path from approach 1 (Petrova Street - West) to approach 3 (Petrova Street - East) was lower by 32.76% that the calculated average design speed. An average vehicle speed measured at the Voćarska - Bijenička roundabout on the path from approach 1 (N. Grškovića Street) to approach 3 (Voćarska Street) was lower by 31.05% that the calculated average design speed, while for the path from approach 2 (Mesićeva Street) to approach 4 (Bijenička Street) was greater by 16.30% that the calculated average design speed. An average vehicle speed measured at the Radnička – Petruševec 1 roundabout on the path from approach 3 (Radnička cesta – South) to approach 1 (Radnička cesta – North) was lower by 38.91% that the calculated average design speed.

Deviation of average measured speeds from average calculated speeds could be explained as follows. The Sv. Duh – Kuniščak roundabout is located at the transition from the mountainous terrain to a flat zone, and the entire intersection is inclined by 5–7%. Due to space restrictions and the need to accommodate heavy vehicle traffic from approach 1 (Sv. Duh – South) to approach 2 (Kuniščak Street), which is generated by the Zagreb brewery complex located 500 m to the east of the roundabout, the roundabout was realized with a traversable central island. Primary school is located to the east of the roundabout between the approaches 1 and 3 (Sv. Duh – South and North). Taking all this into consideration, as well as the information about the percentage of

Table 2. Calculated average design speed and average vehicle speed measured at selected roundabouts in 2016.

Roundabout/approach name	Average path radii (N = 50) [m]			Slope e [m/m]						
	R_1	R_2	R_3	App. R_1	C. r. R_2	App. R_3	P_{HV} (dec)	f_{SLV}	f_{SHV}	f_S
Sveti Duh - Kuniščak										
Approach 1	28.0	33.0	40.0	0.02	−0.015	0.05	0.11	0.27	0.21	0.26
Approach 3	30.0	35.0	41.0	−0.05	−0.015	−0.02	0.18	0.27	0.21	0.26
Petrova - Jordanovac										
Approach 1	20.0	25.0	31.0	0.00	−0.015	−0.02	0.15	0.27	0.21	0.26
Approach 2	20.0	25.4	40.0	0.02	−0.015	0.04	0.18	0.27	0.21	0.26
Approach 3	21.0	27.5	32.0	0.02	−0.015	0.00	0.17	0.27	0.21	0.26
Approach 4	35.0	42.0	44.0	−0.04	−0.015	−0.02	0.04	0.27	0.21	0.27
Voćarska - Bijenička										
Approach 1	28.0	37.0	42.0	0.05	−0.03	0.01	0.12	0.27	0.21	0.26
Approach 2	15.0	20.0	31.0	0.01	−0.03	0.03	0.07	0.27	0.21	0.26
Approach 3	15.0	21.0	24.0	0.02	−0.03	−0.03	0.03	0.27	0.21	0.27
Approach 4	22.0	24.0	29.0	−0.02	−0.03	−0.01	0.05	0.27	0.21	0.27
Radnička road – Petruševec 1										
Approach 1	40.0	45.0	52.0	0.00	−0.005	0.01	0.65	0.27	0.20	0.22
Approach 2	41.0	47.0	54.6	0.01	−0.005	−0.02	0.55	0.27	0.20	0.23
Approach 3	39.0	42.4	51.0	−0.02	−0.005	0.00	0.68	0.27	0.20	0.22
Approach 4	30.0	36.0	43.0	0.02	−0.005	−0.01	0.51	0.27	0.20	0.23

(*continued*)

Table 2. *(continued)*

Direction of travel	Calculated average speed [km/h]			Measured average speed [km/h]			Deviation from design speed [%]		
Approach	V_1	V_2	V_3	V_1	V_2	V_3	V_1	V_2	V_3
Direction of travel	Calculated average speed [km/h]			Measured average speed [km/h]			Deviation from design speed [%]		
Approach	V_1	V_2	V_3	V_1	V_2	V_3	V_1	V_2	V_3
Sveti Duh - Kuniščak									
1–3	31.69	32.20	39.83	18.33	25.02	31.23	−42.15	−22.29	−21.60
3–1	28.18	32.89	35.23	18.22	25.31	27.01	−35.34	−23.05	−23.33
Petrova - Jordanovac									
1–3	25.63	27.81	30.65	17.23	24.01	29.58	−32.76	−13.66	−3.48
3–1	26.51	27.93	38.83	18.24	25.00	28.33	−31.21	−10.48	−27.04
2–4	27.20	29.09	32.34	21.03	25.32	29.66	−22.68	−12.97	−8.29
4–2	33.01	36.52	37.01	28.33	27.15	31.01	−14.17	−25.66	−16.21
Voćarska - Bijenička									
1–3	32.21	33.00	38.07	22.21	25.69	31.54	−31.05	−22.15	−17.16
3–1	22.87	24.41	33.77	22.89	25.55	32.21	0.08	4.67	−4.61
2–4	23.38	25.13	26.87	22.36	25.88	31.25	−4.35	2.97	16.30
4–2	26.20	26.81	30.69	23.59	26.31	33.59	−9.97	−1.85	9.45
Radnička road – Petruševec 1									
1–3	33.56	35.19	39.12	22.36	28.36	37.58	−33.38	−19.42	−3.94
3–1	35.26	36.55	38.05	21.54	28.69	38.56	−38.91	−21.50	1.34
2–4	31.44	34.00	37.72	24.25	32.11	35.26	−22.88	−5.55	−6.51
4–2	30.96	32.19	34.79	26.00	30.21	34.22	−16.02	−6.14	−1.63

Legend: App. - approach, C. r. - circulatory roadway, P_{HV} - percentage of heavy vehicles in traffic flow [decimal], M_{vLV} - average weight of light vehicles (1400–1500) [kg], M_{vHV} - average weight of heavy vehicles (11000–15000) [kg], f_{SLV} - coefficient of friction for light vehicles, f_{SHV} - coefficient of friction for heavy vehicles, f_s - coefficient of friction for vehicles.

heavy vehicles in the total traffic (11%), we can explain the –42.15% deviation of the calculated speed from the average measured speed.

The Petrova – Jordanovac roundabout is located at the foot of a hillside (approach 4 (Jordanovac – North) inclined by 4%) while other approaches and the circulatory roadway are not characterized by greater terrain limitations. Due to restricted space, design elements are smaller and this greatly affects roundabout TE. Primary school is located to the east of the intersection between approaches 2 and 3 (Jordanovac – South and Petrova – East). Having this in mind, and the data about the percentage of heavy vehicles in the total traffic (13.5%, including public transport vehicles), provide proper explanation for the −32.76% deviation of the calculated speed from the measured speed.

The Voćarska - Bijenička roundabout is located at the transition from the hillside to the flat terrain, and the entire roundabout is realized at the grade of 3%, while the grade at the approach 1 (N. Grškovića) amounts to 5%. Due to restricted space, design elements are smaller and this greatly affects roundabout TE. Primary school is located to the south of the roundabout, while the Faculty of Science, Institute for Physics, and the Ruđer Bošković Institute, are located some 300 m to the north. The positioning of these institutions calls for a greater intensity of public transport traffic, as can be seen from the data on the proportion of heavy vehicles in the total traffic (6.75%). The above information suitably explains the −31.05% deviation of the calculated speed from the measured speed.

The Radnička – Petruševec 1 roundabout is situated in a flat area, but is located at the heavily trafficked Radnička street in the south-eastern part of Zagreb. The roundabout was built to properly link the south-eastern part of Zagreb, i.e. the nearby industrial zone situated 500 m to the north of the roundabout, with the Zagreb Bypass (Kosnica Interchange), and the Pleso Airport, via the Homeland Bridge. Primary school is located to the west of the intersection between approaches 2 and 3 (Petruševec 1 and Radnička – South). The above information, and the data about the percentage of heavy vehicles in the total traffic (59.75%, including public transport vehicles), provide proper explanation for the −38.91% deviation of the calculated speed from the measured speed.

3.2 Results Comparison

Here the studied results from previous work [14], and the 2016 measurements will be briefly presented (Table 3). Only total average data deviations for all four roundabouts will be shown. The latest *in situ* measurements of average vehicle speed are greater from 4 to 8% in total comparing the measurements from 2011. Regarding the average increase of percentage of heavy vehicle of 28.4%, the calculated average design speeds remain the same (from 29 to 35 km/h). Deviations of measured and calculated design speeds are for 2016 smaller than for 2011, by; V1 (−28%), V2 (−126%), and V3 (−65%). The biggest deviations of measured versus calculated design is registered on Petrova – Jordanovac roundabout; approach 3 to 1 has decrease of 1256.8%, and approach 2 to 4 increase of 24.74%.

Table 3. Deviation between 2011 and 2016 data measurements.

Direction of travel	Deviation in 2011 [14] [%]			Deviation in 2016 [%]			Deviation [%]		
Roundabout Approach	V_1	V_2	V_3	V_1	V_2	V_3	V_1	V_2	V_3
Sveti Duh - Kuniščak									
1–3	−44.4	−25.0	−26.0	−42.1	−22.2	−21.6	−5.2	−10.8	−17.2
3–1	−42.8	−25.5	−24.0	−35.3	−23.0	−23.3	−17.4	−9.7	−2.7
Petrova - Jordanovac									
1–3	−34.1	−16.3	−9.1	−32.7	−13.6	−3.4	−4.1	−16.4	−61.8
3–1	−38.1	−17.1	−34.5	−31.2	−10.4	−27.0	−18.1	−39.0	−21.8
2–4	−25.3	−13.4	−11.5	−22.6	−12.9	−8.2	−10.5	−3.8	−28.3
4–2	−22.9	−27.1	−19.4	−14.1	−25.6	−16.2	−38.2	−5.5	−16.5
Voćarska - Bijenička									
1–3	−38.8	−24.5	−20.6	−31.0	−22.1	−17.1	−20.1	−9.8	−16.6
3–1	−7.6	−0.4	−14.9	0.08	4.6	−4.6	−101.0	−1256.8	−69.0
2–4	−7.3	−2.6	13.0	−4.3	2.9	16.3	−41.1	−211.3	24.7
4–2	−18.4	−3.2	−3.0	−9.9	−1.8	9.4	−46.0	−42.9	−410.3
Radnička road – Petruševec 1									
1–3	−38.9	−24.0	−12.9	−33.3	−19.4	−3.9	−14.2	−19.1	−69.5
3–1	−45.6	−30.4	−17.8	−38.9	−21.5	1.3	−14.8	−29.2	−107.5
2–4	−28.9	−13.9	−10.4	−22.8	−5.5	−6.5	−20.9	−60.2	−37.4
4–2	−26.9	−13.8	−7.7	−16.0	−6.1	−1.6	−40.4	−55.7	−78.9

4 Discussion and Conclusion

Implementing safe and effective roundabouts in urban areas is a highly demanding task and requires optimizing TE and TS while considering geometric factors, traffic characteristics and local constraints. In designing urban single-lane roundabouts a special attention must be paid to the design of individual roundabout elements (i.e. ICD, CID, width of circulatory roadway and approach lanes) to comply with the vehicle path speed needed for proper negotiation of the roundabout. Determination of a method for defining design speed for safe negotiation of the roundabout geometry is needed. This is also for better understanding TE and TS parameters, both at the new roundabout modelling phase, and during analysis of existing roundabouts.

In this paper we repeated the study of single-lane roundabout vehicle design speed using the same methodology as we used in [14]. In general, it can be concluded that the basic roundabout design requirement of R_1, $R_2 < R_3$, according to [16], has been met. Deviations between the average design speeds and measured speeds vary from −42.15% to +16.30% for all studied roundabouts. The latest *in situ* measurements of average vehicle speed are greater from 4 to 8% comparing the measurements from 2011. Regarding the average increase of percentage of heavy vehicle of 28.4%, the calculated average design speeds remain the same (from 29 to 35 km/h). Deviations of measured and calculated design speeds are for 2016 smaller than for 2011, by;

V_1 (−28%), V_2 (−126%), and V_3 (−65%). These deviations result from: used design speed methods, location features of intersections and their disposition within the city traffic network, roundabout area design elements and traffic equipment, various axle loads and properties of vehicles, traffic flow characteristics, and driver behavior and experience.

Limitations of the study reflects the same limitations that are emphasized in our previous study [14]; we did not study path radii (R_4 and R_5, i.e. (V_4) and (V_5)); sample size - 50 measurements for vehicle path radii and design speed on four single-lane roundabouts; and we did not study the influence of design elements on the vehicle path speed (i.e. on the TE and TS).

Future study should involve a greater number of similar roundabouts with greater size samples of vehicle path radii and design speed, and vehicle path speed for left/right turns at roundabouts. It could also be necessary to study and correlate the driver behavior and type of vehicles with vehicle speed during measurement. In addition, it would be useful to calibrate the used methods to local conditions. If the used methods can be validated, it may be possible to expand it to include several parameters, pedestrian and bicyclist demand flow. In addition, future studies should examine this approach on other urban/suburban single and multi-lane roundabout types - turbo roundabouts [20–22]. Development and validation of such methods would be useful to better understand the TE and TS parameters for Croatian conditions or other. Therefore, further work is needed to confirm the validity of our approach.

Acknowledgement. The research described in this paper was conducted within the scope of the research project "Correlation between Design and Safety at Roundabouts" (135-0000000-3313), funded by the Ministry of Science, Education and Sports of the Republic of Croatia. The *in situ* studies and data analysis was conducted by the student Šime Šarić within the scope of his master thesis "A Study of the Impact of Roundabouts on Functional Efficiency of Intersections and Business" at University of Zagreb, Faculty of Transport and Traffic Safety, Croatia.

References

1. Vasconcelos, A., Seco, A., Silva, A.: Comparison of procedures to estimate critical headways at roundabouts. Promet **25**, 43–53 (2013)
2. Mauro, R., Cattani, M.: Functional and economic evaluations for choosing road intersection layout. Promet **24**, 441–448 (2012)
3. Kim, S., Choi, J.: Safety analysis of roundabout designs based on geometric and speed characteristics. J. Civ. Eng. **17**, 1446–1454 (2013)
4. Rubio-Martín, J.L., Jurado-Piña, R., Pardillo-Mayora, J.M.: Heuristic procedure for the optimization of speed consistency in the geometric design of single-lane roundabouts. Can. J. Civ. Eng. **42**, 13–21 (2015)
5. Vujanić, M., Savićević, M., Antić, B., Pešić, D.: Safety effectiveness of converting conventional intersections to roundabouts: case study in the city of Niš. Promet **28**, 529–537 (2016)
6. Macioszek, E.: Geometrical determinants of car equivalents for heavy vehicles crossing circular intersections. In: Mikulski, J. (ed.) Telematics in the Transport Environment. CCIS, vol. 329, pp. 221–228. Springer, Heidelberg (2012)

7. Zirkel, B., Park, S., McFadden, J., Angelastro, M., McFarthy, L.: Analysis of sight distance, crash rate and operating speed relationships for low-volume single lane roundabouts in the United States. J. Transp. Eng. **139**, 565–573 (2013)
8. Quddus, M., Washington, S.: Shortest path and vehicle trajectory aided map-matching for low frequency GPS data. Transp. Res. Part C **55**, 328–339 (2015)
9. Deluka-Tibljaš, A., et al.: Smjernice za Projektiranje Kružnih Raskrižja na Državnim Cestama. University of Rijeka, Faculty of Civil Engineering, Rijeka (2014)
10. BiH, Direkcija Cesta Federacije: Javno Preduzeće Putevi Republike Srpske: Smjernice za Projektovanje, Građenje. Održavanje i Nadzor nad Putevima. Knjiga I. Direkcija Cesta Federacije BiH, Sarajevo/Banja Luka (2005)
11. Ministarstvo za Infrastrukturo in Prostor DRS za Ceste: Krožna Križišča. Ministarstvo za Infrastrukturo, Ljubljana (2011)
12. FGSV: Handbuch für die Bemessung von Strassenverkehrsanlagen: HBS 2015. FGSV, Köln (2015)
13. Generalna Dyrekcja Dróg Publicznych: Wytyczne Projektowania Skrzyżowań Drogowych. Część II – Ronda. GDDP, Warszawa (2001)
14. Pilko, H., Brčić, D., Šubić, N.: Study of vehicle speed in the design of roundabouts. Građevinar **66**, 407–416 (2014)
15. Pilko, H.: Optimization of Roundabout Design and Safety Component. University of Zagreb, Zagreb (2014)
16. Transportation Research Board: NCHRP Report 672. National Cooperative Highway Research Program. Roundabouts: An Informational Guide. Transportation Research Board, Washington (2010)
17. Easa, S.M., Mehmood, A.: Optimizing geometric design of single-lane roundabouts: consistency analysis. Can. J. Civ. Eng. **31**, 1024–1038 (2004)
18. Legac, I., et al.: Korelacija Oblikovnosti i Sigurnosti u Raskrižjima s Kružnim Tokom Prometa. Ministry of Science, Education and Sports, Zagreb (2008)
19. Sabor, Hrvatski: Pravilnik o Tehničkim Uvjetima Vozila u Prometu na Cestama, NN 85/2016. Hrvatski Sabor, Zagreb (2016)
20. Macioszek, E.: Analysis of significance of differences between psychotechnical parameters for drivers at the entries to one-lane and turbo roundabouts in poland. In: Sierpiński, G. (ed.) Intelligent Transport Systems and Travel Behaviour. AISC, vol. 505, pp. 149–161. Springer, Switzerland (2017)
21. Džambas, T., Ahac, S., Dragčević, V.: Design of turbo roundabouts based on the rules of vehicle movement geometry. J. Transp. Eng. **143**, 1–10 (2016)
22. Hatami, H., Aghayan, I.: Traffic efficiency evaluation of elliptical roundabout compared with modern and turbo roundabouts considering traffic signal control. Promet **29**, 1–11 (2017)

The Comparison of Models for Follow-up Headway at Roundabouts

Elżbieta Macioszek(✉)

Faculty of Transport, Silesian University of Technology, Katowice, Poland
elzbieta.macioszek@polsl.pl

Abstract. One of the basic parameter used for roundabout entry capacity calculation is follow-up headway. This parameter beyond critical headway and traffic volume on roundabout roadway major for vehicle drivers from entry determines roundabout entry capacity. The comparative analysis of models and studies on follow-up headway estimation have been presented in this article. The analysis were made for single lane, two lane and turbo roundabouts. The largest number of models and studies of follow-up headway parameter refers to single lane roundabouts. The least studies were devoted to values of follow-up headway parameter at turbo roundabouts due to the fact that this is still relatively a new type of roundabout in compare with single lane or two lane roundabouts.

Keywords: Roundabouts · Follow-up headway · Gap acceptance theory · Road traffic engineering

1 Introduction

In recent years roundabouts are very popular kind of intersection. They are subject of multifaceted scientific research concerning both considerations on microscopic level f.ex. like their performance, capacity, safety (f.ex.: [1–10]) and also considerations on macroscopic level as element of the transport systems [11–19]. Due to the high level of traffic safety roundabouts are willingly used as a part of transport networks in urban area and outside urban area. The roundabouts popularity resulted in the development of models and methods for mapping behavior of traffic streams (vehicle driver, pedestrians, cyclists) on roundabouts (f.ex: [20–29]).

In models intended to roundabout entry capacity calculations based on gap acceptance theory the critical headway and follow-up headway are the basic parameters in vehicle drivers decision-making process at roundabout entry. These parameters usually represent the average behaviors in vehicle drivers population and together with the headway distribution between vehicles on main roadway of roundabout decides about the degree of accuracy of roundabout entry capacity determination.

According to literature [30] the follow-up headway (t_f) is defined as the time between the departure of first (previous) vehicle from the roundabout entry and the departure of the next vehicle using the same major-street headway under a condition of queuing on the roundabout entry. If the distance between vehicles in traffic stream on main roadway of roundabout allows for entry of further vehicles from the queue, they are passing through the edge of the roundabout roadway at follow-up headway one

E. Macioszek and G. Sierpiński (eds.), *Recent Advances in Traffic Engineering for Transport Networks and Systems*, Lecture Notes in Networks and Systems 21,
DOI 10.1007/978-3-319-64084-6_2

after another. At very low traffic volume on main roadway of roundabout the value of follow-up headway mainly decides about the entry capacity. According to Polish recommendations [30] setting for the roundabout capacity calculation the follow-up headway values the type of roundabout, roundabout external diameter and number of entry lanes are taking account.

In case of assumption that on roundabout entry is queue of vehicles and vehicle drivers from this queue use the same gap between vehicles moving on main roadway of roundabout the follow-up headway value can be calculated from equation:

$$t_f = t_f^{next} - t_f^{previous} \ [s] \tag{1}$$

where:

t_f - follow-up headway [s],

t_f^{next} - crossing time the edge of the roundabout roadway by next vehicle driver [s],

$t_f^{previous}$ - crossing time the edge of the roundabout roadway by previous vehicle driver [s].

Many scientific studies provide results from measurements of follow-up headway from real data at roundabouts. In practice, the value of parameter t_f is frequently represented as the average value (specifying also with value of the standard deviation) or as a median. In different scientific papers the minimum and maximum values are also presented. In case of two lane roundabouts the follow-up headway values are stated separately for the right and left entry lane as well as for the inner and outer lane on roundabout roadway. Whereas in case of turbo roundabouts further decomposition of t_f values occurs because of different types of traffic control on entry and on roundabout main roadway area, i.e. one or two entry lanes and one or two circulating lanes.

The analysis in the regards of scientific papers related to determination of follow-up headway values for vehicle drivers on single lane, two lane and turbo roundabouts have been presented in this article. In view of a very large collection of published studies connected with this issue, the analysis in this article is focused on the most important, most popular models and the latest studies in this area.

2 The Comparison of Models for Follow-up Headway at Roundabouts

In scientific literature can be found a lot of synthetic reviews on gap acceptance parameters for roundabouts, f.ex. [31–35]. According to research the values of follow-up headways may depends on roundabout geometry, number of traffic lanes on entry and on main road of roundabout, vehicle type, the presence of bunched vehicles on main roadway and many other factors. The follow-up headway value also depend on combination of vehicle types in the queue at the entry (i.e. passenger cars and heavy vehicles). In many scientific papers the follow-up headway value is defined for passenger cars, less frequently for another type of vehicle like heavy vehicles, bikes, buses etc. The selected studies characterizing the follow-up headway for drivers at single lane roundabouts entries are presented in Table 1.

Table 1. Comparison of follow-up headway models for single lane roundabouts.

Country	Author	The model/value of follow-up headway parameter [s]
Germany	W. Brilon [36]	For small roundabouts ($13 \leq D_z \leq 26$ m) and single roundabouts ($26 \leq D_z \leq 40$): $t_f = 2.84 + \frac{2.07}{D_z}$, $t_f \in \langle 2.89; 3.00 \rangle$, (Average value = 2.94)
	N. Wu [37]	Average value = 2.88
USA	HCM 2010 [38]	Average value = 3.20
	NCHRP Report 572 [39]	$t_f \in \langle 2.6; 4.3 \rangle$, (Average value = 3.20)
	HCM 2016 [40]	Average value = 2.60
	Y. Mereszczak et al. [41]	$t_f \in \langle 2.6; 3.0 \rangle$, (Average value = 2.80)
	F. Xu, Z. Tian [42]	$t_f = 2.38 - 0.000071 \cdot Q_{nwl}$, (Average value = 2.50)
Canada	J. Dahl, Ch. Lee [43]	$t_f \in \langle 2.10; 4.20 \rangle$, (Average value = 3.15)
The Netherlands	L. Fortuijn [44]	Average value = 2.10
Australia	L. Ren, Q. Xiaobo [45]	$t_f \in \langle 2.35; 2.75 \rangle$, (Average value = 2.55)
	X. Qu [46]	Average value = 2.76
Portugal	L. Vasconcelos, A. Seco, A. Silva [47]	$t_f \in \langle 2.10; 2.30 \rangle$, (Average value = 2.20)
Spain	M. Romana [48]	$t_f \in \langle 1.65; 1.75 \rangle$, (Average value = 1.70)
Denmark	P. Greibe [49]	Average value = 3.00
Italy	A. Gazzarri et al. [50, 51]	$t_f \in \langle 2.52; 2.76 \rangle$, (Average value = 2.63)
Slovenia	T. Tollazi [52]	Average value = 2.90
Poland	J. Chodur [53]	$t_f = 0.31 \cdot D_z - 0.0044 \cdot D_z^2 + 0.00052 \cdot wm - 2.59$ (Average value = 2.85)
	E. Macioszek [54]	For 22 m $\leq D_z \leq 45$ m, 4 $m \leq l_{jr} \leq 8$ m: $t_f = 3.64 - 0.02 \cdot D_z - 0.03 \cdot l_{jr}$ (Average value = 2.79)

where:
D_z - roundabout external diameter [m],
Q_{nwl} - circulating flow [pcu/h],
wm - city size determined by the number of inhabitants (19.60-740.00 thousands of inhabitants),
$\overline{V}$ - the average vehicle speed [km/h],
l_{jr} - the width of the roundabout roadway [m],
t_g - critical headway [s].

On the basis of data included in Table 1 can be concluded the differences between the proposed follow-up headways values. These differences are the results of various vehicle driver behavior in different countries in the world as well as result from the different techniques and methods of follow-up time headway parameter measurements. The comparison of average values of follow-up headways for single lane roundabouts proposed by different authors in different countries of the world have been presented on the Fig. 1.

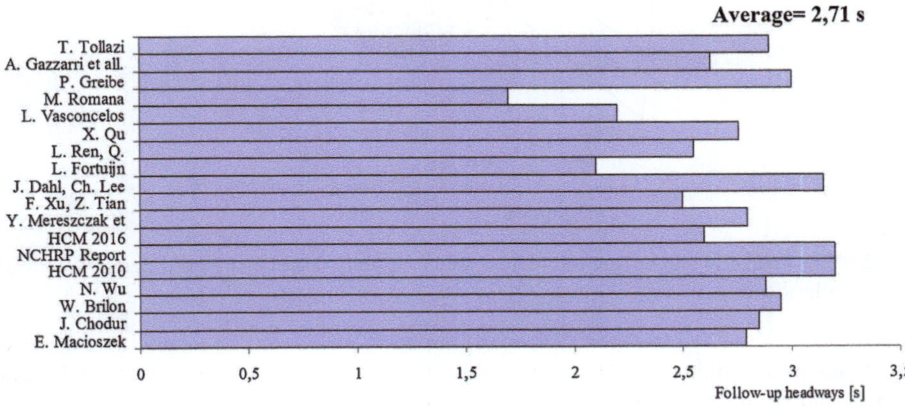

Fig. 1. The average values of follow-up headways for single lane roundabouts

On the basis of Fig. 1 can be concluded that the average value of follow-up headway for all models is equal 2.71 s. Values below the average can be observed f.ex. in such countries as Italy, Spain, Portugal, USA (according HCM 2016), Netherlands, Australia. Whereas values higher than the average value can be observed f.ex. in Germany, Canada, Denmark, Slovenia and Poland. It should be noted that according most recent research published in HCM 2016 average value of follow-up headway is equal 2,60 s, while previous USA studies (f.ex. HCM 2010, NCHRP Report 572, F. Xu, Z. Tian) indicate for a slightly higher average values of follow-up headway for single lane roundabouts. The decrease in average value of follow-up headway confirms that vehicle drivers in USA have become accustomed with roundabouts and they drive firmly on this type of intersections.

Besides, the potential impact of follow-up headway value on single lane roundabout entry capacity was examined by varying the parameter t_f from minimum to maximum values under different circulating traffic flows (Q_{nwl}). The follow-up headway value change between 1.70–3.20 s. This t_f values were adopted on the basis of various authors research presented in Table 1 (see Fig. 2). Critical headway was assumed to be constant. As can be seen from Fig. 2 single lane roundabout entry capacity is higher when vehicle drivers from queue at roundabout entry moving on main roadway of roundabout with smaller follow-up headway values. Consequently, accurate determination of follow-up headway value is very important because can improves the accuracy of roundabout entry capacity calculation.

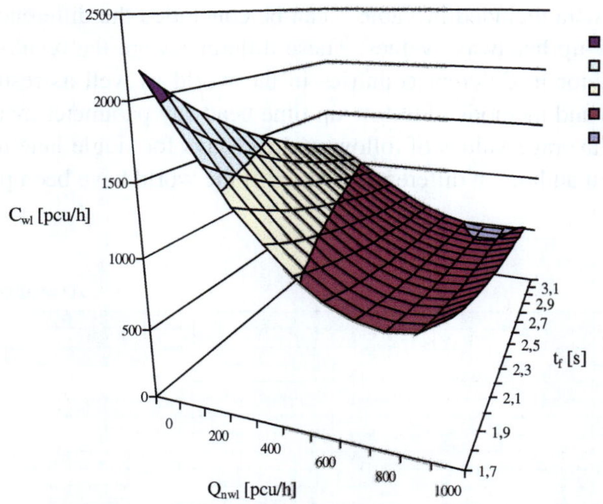

Fig. 2. The change in single lane roundabout entry capacity calculated with different values of follow-up headways

In scientific studies can be found slightly less number of follow-up headways models and results for two lane roundabouts then in case of single lane roundabouts. The comparison of selected follow-up headway results for two lane roundabout have been presented in Table 2. Base on the Table 2 can be concluded that part of these results were defined with accuracy of roundabout entry lane and other part applies only to value of follow-up headway for entire roundabout entry.

The comparison of average values of follow-up headways for two lane roundabouts proposed by different authors in different countries of the world have been presented on the Fig. 3. Based on the Fig. 3 can be concluded that average values of follow-up headway for vehicle drivers from right and left entry lane are respectively 2.68 s and 2.72 s. It means that vehicle drivers from the right entry lane moving on main roadway of roundabout faster, with shorter follow-up headways then vehicle drivers from the left entry lane.

One of new types of roundabouts like turbo roundabouts were already development and implemented in many countries in the world. In case of turbo roundabouts the follow-up headway value depend on traffic control in entry and roundabout main roadway area. In practice at turbo roundabouts can be distinguish few different type of traffic control in entry and roundabout main roadway area i.e. one or two entry lane and one or two circulating lane. The comparison of follow-up headway values for turbo roundabouts have been presented in Table 3. Due to the fact that turbo roundabouts are relatively new type of intersection, there are little research results and models for follow-up headways. There is a definite need for further research on follow-up headways values at turbo roundabouts at different variants of traffic control.

Table 2. Comparison of follow-up headway results for two lane roundabouts.

Country	Author	The value of follow-up headway parameter [s]
Germany	W. Brilon [36]	$t_f \in \langle 2.20; 2.90 \rangle$, (Average value = 2.55),
	N. Wu [37]	Average value = 2.88
USA	HCM 2010 [38]	Left lane average = 3.20, Right lane average = 3.20
	HCM 2016 [40]	Left lane average = 2.66, Right lane average = 2.53
	F. Xu, Z. Tian [42]	Left lane average = 2.30, Right lane average = 2.20
	NCHRP Report 572 [39]	Left lane = 3.1-4.7, (Average value = 3.4), Right lane = 2.7-4.4, (Average value = 3.1)
Canada	J. Dahl, Ch. Lee [43]	$t_f \in \langle 1.60; 5.00 \rangle$, (Average value = 3.30)
Portugal	L. Vasconcelos, A. Seco, A. Silva [47]	$t_f \in \langle 1.94; 2.78 \rangle$, (Average value = 2.36)
Switzerland	N. Leemann, G. Santel [55]	$t_f \in \langle 2.27; 2.63 \rangle$, (Average value = 2.45)
Denmark	P. Greibe [49]	Average value = 2.60
	O. Hagring, N. Rouphail, H. Sorensen [56]	Average value = 2.79
Italy	A. Gazzarri, M. Martello, A. Pratelli, R. Souleyrette [50]	Left lane average = 2.65, Right lane average = 2.64, Minimum value = 2.16, Maximum value = 3.10
The Netherlands	L. Fortuijn [44]	$t_f \in \langle 2.24; 2.26 \rangle$, (Average value = 2.25)
Poland	Polish Quidelines [30]	$t_f \in \langle 2.90; 3.30 \rangle$, (Average value = 3.10)
	E. Macioszek [54]	Left lane average = 3.20, Right lane average = 3.20

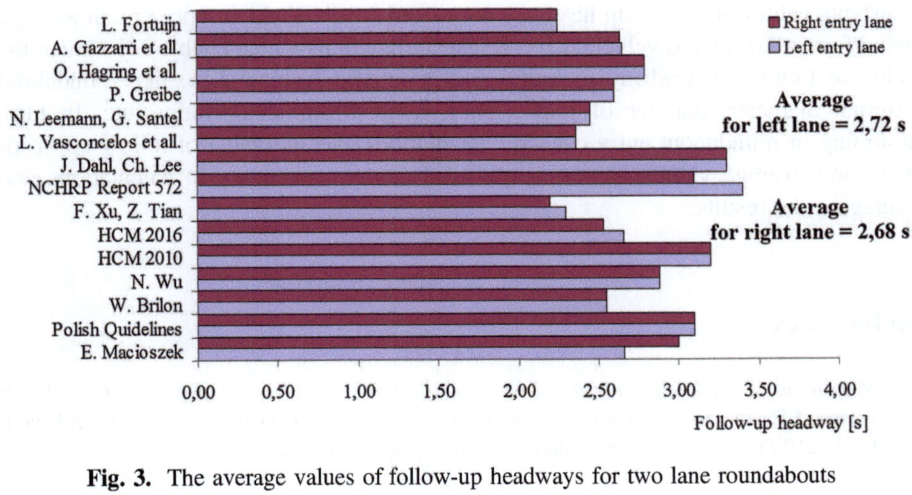

Fig. 3. The average values of follow-up headways for two lane roundabouts

Table 3. Comparison of follow-up headway results for turbo roundabouts.

Country	Author	The value of follow-up headway parameter [s]
The Netherlands	L. Fortuijn [44]	Right entry lane at major entry: average 2.13 Left entry lane at major entry: average 2.25 Depending on traffic control
Germany	W. Brilon, L. Bondzio, F. Weiser [57]	$t_f \in \langle 2.50; 2.80 \rangle$ Depending on traffic control
Poland	E. Macioszek [54]	Right lane: 1.48-4.76 (average 2.21) Left lane: 1.24–5.08 (average 2.35) Depending on traffic control

3 Summary

The synthesis from scientific studies about follow-up headway values for single lane, two-lane and turbo roundabouts have been presented in this article. The largest number of models and studies of t_f parameter refers to single lane roundabouts. This is connected with a large number of such type of roundabouts developed in different countries of the world. Because of their great advantages in the context of road traffic safety in many countries single lane roundabouts are the most numerous group among roundabouts. By far the least studies were devoted to values of t_f parameter at turbo roundabouts due to the fact that this is still relatively a new type of roundabout in compare with single lane or two lane roundabouts. There is a definite need for further research on follow-up headways values at turbo roundabouts at different variants of traffic control.

The values of follow-up headway for particular types of roundabouts are different depending on country and individual studies. These differences have an influence on calculated roundabout entry capacity. In addition, in many studies in calculations the constants values of follow-up headway are recommended (which represent an average value from all observed vehicle drivers) instead functions which enable adjustment the value of follow-up headway to traffic and geometric features f.ex. like roundabout external diameter, number of traffic lanes, lanes width, entry radius etc. In turn, assuming in roundabout entry capacity calculations average value of follow-up headway (and average value of critical headway) causes inaccuracy in calculations and averaging the results.

References

1. Macioszek, E., Akçelik, R.: A comparison of two roundabout capacity models. Paper Prepared for the 5th International Roundabout Conference. Transportation Research Board, USA (2017). http://www.sidrasolutions.com/Resources/Articles

2. Macioszek, E.: Relationship between vehicle stream in the circular roadway of a one-lane roundabout and traffic volume on the roundabout at peak hour. In: Mikulski, J. (ed.) Telematics - Support for Transport. CCIS, vol. 471, pp. 110–119. Springer, Heidelberg (2014)
3. Macioszek, E.: Turbo-roundabouts signing and marking – current situation in Poland. Arch. Transp. Syst. Telemat. **6**(1), 26–30 (2013)
4. Szczuraszek, T., Macioszek, E.: Analiza Rozkładów Odstępów Czasu Pomiędzy Pojazdami na Obwiedni Małych Rond. Drogi i Mosty **3**, 89–99 (2010)
5. Macioszek, E.: Analysis of the effect of congestion in the lanes at the inlet to the two-lane roundabout on traffic capacity of the inlet. In: Mikulski, J. (ed.) Activities of Transport Telematics. CCIS, vol. 395, pp. 97–104. Springer, Heidelberg (2013)
6. Macioszek, E.: The influence of motorcycling and cycling on small one-lane roundabouts capacity. In: Mikulski, J. (ed.) Modern Transport Telematics. CCIS, vol. 239, pp. 291–298. Springer, Heidelberg (2011)
7. Macioszek, E.: The application of HCM 2010 in the determination of capacity of traffic lanes at turbo roundabout entries. Transp. Probl. **11**(3), 77–89 (2016)
8. Macioszek, E.: Analiza Możliwości Zastosowania Metody HCM 2010 do Szacowania Przepustowości Pasów Ruchu na Wlotach Rond w Polsce. Prace Nauk. Politech. Warsz. Ser. Transp. **113**, 327–340 (2016)
9. Macioszek, E.: The movement model for small roundabouts with minor roads capacity estimating. Transp. Probl. **1**, 87–94 (2007)
10. Macioszek, E.: Analysis of opportunities of using the HCM 2010 methodology for evaluation of traffic capacity at the entries to turbo roundabouts. In: Proceedings VIII International Scientific Conference Transport Problems, pp. 314–326 (2016)
11. Sierpiński, G., Staniek, M., Celiński, I.: Research and shaping transport systems with multimodal travels – methodological remarks under the green travelling project. In: 7th International Conference of Education, Research and Innovation ICERI Proceedings, pp. 3101–3107. Seville (2014)
12. Okraszewska, R., Nosal, K., Sierpiński, G.: The role of the polish universities in shaping a new mobility culture - assumptions, conditions, experience. Case study of Gdansk University of Technology, Cracow University of Technology and Silesian University of Technology. In: 7th International Conference of Education, Research and Innovation ICERI Proceedings, pp. 2971–2979. Seville (2014)
13. Sierpiński, G.: Theoretical model and activities to change the modal split of traffic. In: Mikulski J. (ed.) Transport Systems Telematics 2012. Telematics in the Transport Environment. CCIS, vol. 329, pp. 45–51. Springer, Heidelberg (2012)
14. Staniek, M.: Road pavement condition as a determinant of travelling comfort. In: Sierpiński, G. (ed.) Intelligent Transport Systems and Travel Behaviour. AISC, vol. 505, pp. 99–107. Springer, Switzerland (2017)
15. Sierpinski, G.; Staniek, M., Celinski, I.: New methods for pro-ecological travel behavior learning. In: 8th International Conference of Education, Research and Innovation ICERI Proceedings, pp. 6926–6933. Seville (2015)
16. Staniek, M.: Stereo vision method application to road inspection. Balt. J. Road Bridge Eng. **12**(1), 38–47 (2017)
17. Burdzik, R., Celiński, I., Czech, P.: Optimization of transportation of unexploded ordnance, explosives and hazardous substances - vibration issues. Vibroeng. Proc. **10**, 382–386 (2016)

18. Młyńczak, J., Burdzik, R., Celiński, I.: Research on dynamics of shunting locomotive during movement on marshalling yard by using prototype of remote control unit. In: Awrejcewicz, J. (ed.) Dynamical Systems: Theoretical and Experimental Analysis. PMS, vol. 182, pp. 279–292. Springer, Berlin (2016)
19. Celiński, I.: Transport network parameterisation using the GTAlg tool. In: Macioszek, E., Sierpiński, G. (eds.) Contemporary Challenges of Transport Systems and Traffic Engineering. LNNS, vol. 2, pp. 111–123. Springer, Switzerland (2017)
20. Macioszek, E.: Geometrical determinants of car equivalents for heavy vehicles crossing circular intersections. In: Mikulski, J. (ed.) Telematics in the Transport Environment. CCIS, vol. 329, pp. 221–228. Springer, Heidelberg (2012)
21. Szczuraszek, T., Macioszek, E.: Proportion of vehicles moving freely depending on traffic volume and proportion of trucks and buses. Balt. J. Road Bridge Eng. **8**(2), 133–141 (2013)
22. Macioszek, E., Sierpiński, G., Czapkowski, L.: Methods of modeling the bicycle traffic flows on the roundabouts. In: Mikulski, J. (ed.) Transport Systems Telematics. CCIS, vol. 104, pp. 115–124. Springer, Heidelberg (2010)
23. Macioszek, E., Sierpiński, G., Czapkowski, L.: Problems and issues with running the cycle traffic through the roundabouts. In: Mikulski, J. (ed.) Transport Systems Telematics. CCIS, vol. 104, pp. 107–114. Springer, Heidelberg (2010)
24. Macioszek, E.: Safe road traffic on roundabouts as an element assisting efficient road transportation system development in the Upper Silesia Region. In: Janecki, R., Krawiec, S., Sierpiński, G. (eds.) Contemporary Transportation Systems. Selected Theoretical and Practical Problems. The Transportation as the Factor of the Socio-economic Development of the Regions. Monograph, vol. 386, pp. 85–95. Silesian University of Technology, Gliwice (2012)
25. Macioszek, E.: Road traffic safety as an element creating mobility culture in cities. In: Janecki, R., Sierpiński, G. (eds.) Contemporary Transportation Systems. Selected Theoretical and Practical Problems. New mobility Culture. Monograph, vol. 324, pp. 61–71. Silesian University of Technology, Gliwice (2011)
26. Macioszek, E.: The road safety at turbo roundabouts in Poland. Arch. Transp. **33**(1), 57–67 (2015)
27. Macioszek, E.: Stan Bezpieczeństwa Ruchu Drogowego na Rondach Turbinowych w Polsce. Prace Naukowe Politechniki Warszawskiej. Transport. Bezpieczeństwo i Analiza Ryzyka w Transporcie, vol. 96, pp. 295–307 (2013)
28. Macioszek, E.: The passenger car equivalent factor for heavy vehicles on roundabouts. In: Janecki, R., Sierpiński, G. (eds.) Contemporary Transportation Systems. Selected Theoretical and Practical Problems. The Development of Transportation Systems. Monograph, vol. 256, pp. 127–137. Silesian University of Technology, Gliwice (2010)
29. Macioszek, E.: Analiza Prędkości Przejazdu Pojazdów Przez Skrzyżowania z Ruchem Okrężnym. Prace Naukowe Politechniki Warszawskiej. Transport. Systemy, Podsystemy i Środki w Transporcie Drogowym, Morskim i Śródlądowym, vol. 82, pp. 69–84 (2012)
30. Generalna Dyrekcja Dróg Krajowych i Autostrad: Metoda Obliczania Przepustowości Rond. Instrukcja Obliczania. GDDKiA, Warszawa (2004)
31. Giuffre, O., Grana, A., Tumminello, A.L.: Gap-acceptance parameters for roundabouts: a systematic review. Eur. Transp. Res. Rev **8**, 1–20 (2016)
32. Macioszek, E.: Analysis of significance of differences between psychotechnical parameters for drivers at the entries to one-lane and turbo roundabouts in Poland. In: Sierpiński, G. (ed.) Intelligent Transport Systems and Travel Behaviour. AISC, vol. 505, pp. 149–161. Springer, Switzerland (2017)
33. Macioszek, E., Woch, J.: The Follow-up time issue on small roundabouts. Transp. Probl. **3**, 25–31 (2008)

34. Macioszek, E.: The comparison of models for critical headways estimation at roundabouts. In: Macioszek, E., Sierpiński, G. (eds.) Contemporary Challenges of Transport Systems and Traffic Engineering. LNNS, vol. 2, pp. 205–219. Springer, Switzerland (2017)
35. Macioszek, E., Sierpiński, G. (eds.) Contemporary challenges of transport systems and traffic engineering. Selected Papers. LNNS, vol. 2. Springer, Switzerland (2017)
36. Brilon, W.: Studies on Roundabouts in Germany. Lessons Learned. http://www.ruhr-uni-bochum.de/verkehrswesen/download/literatur/Brilon_roundabouts_2011_05_29_cit.pdf
37. Wu, N.: An Universal Formula for Calculating Capacity at Roundabouts. Research Report of Institute for Traffic Engineering. Ruhr University, Bochum (1997)
38. Transportation Research Board: Highway Capacity Manual 2010. Transportation Research Board of the National Academy of Science, Washington (2010)
39. Rodegerdts, L., Blogg, M., Wemple, E., Myers, E., et al.: NCHRP Report 572. Roundabouts in the United States. Transportation Research Board, Washington (2007)
40. Transportation Research Board: Highway Capacity Manual 6th Edition. A Guide for Multimodal Mobility Analysis. Transportation Research Board of the National Academy of Science, Washington (2016)
41. Mereszczak, Y., Dixon, M., Kyte, M., Rodegerdts, L.A., Blogg, M.: Incorporating exiting vehicles in capacity estimation at single-lane U.S. roundabouts. http://citeseerx.ist.psu.edu/viewdoc/download?doi=10.1.1.508.7303&rep=rep1&type=pdf
42. Xu, F., Tian, Z.Z.: Driver behavior and gap acceptance characteristics at roundabouts in California. J. Transp. Res. Board **2071**, 117–124 (2008)
43. Dahl, J., Lee, C.: Empirical estimation of capacity for roundabouts using adjusted gap-acceptance parameters for trucks. J. Transp. Res. Board **2312**, 34–45 (2012)
44. Fortuijn, L.G.H.: Turbo roundabout. Estimation of capacity. Transp. Res. Rec. **2130**, 83–92 (2009)
45. Ren, L., Xiaobo, Q.: Impact analysis of headway distributions on single-lane roundabout entry capacity. J. Eng. Proj. Production Manag. **6**(1), 2–12 (2016)
46. Qu, X., Zhang, I.J., Wang, S., Liu, Z.: Modelling follow up time at a single lane roundabout. J. Traffic Transp. Eng. **1**, 97–102 (2014)
47. Vasconcelos, A.L.P., Seco, A.J.M., Silva, A.M.C.B.: Comparison of procedures to estimate critical headways at roundabouts. Promet Traffic Transp. **25**, 43–53 (2013)
48. Romana, M.G.: Roundabout practice and research in Spain. In: 6th International Symposium on Highway Capacity (2011)
49. Greibe, P.: Roundabout - design and capacity. In: Denmark Country report, 6th International Symposium on Highway Capacity (2011)
50. Gazzarri, A., Martello, M., Pratelli, A., Souleyrette, R.: Estimation of gap acceptance parameters for HCM 2010 roundabout capacity model applications. In: Longhurst, J.W.S., Brebbia, C.A. (eds.) Urban Transport XVIII. Urban Transport and the Environment in the 21st Century, vol. 128, pp. 309–320. WIT Press, Southampton (2012)
51. Gazzarri, A., Martello, M., Pratelli, A., Souleyrette, R.: Gap acceptance parameters for HCM 2010 roundabout capacity model applications in Italy. Civil Eng. Fac. Publ. **1**, 1–16 (2013)
52. Tollazi, T.: The contribution to the procedure of capacity determination at unsignalized prioriry controlled intersections. Promet Traffic Transp. **16**, 31–36 (2004)
53. Chodur, J.: Funkcjonowanie Skrzyżowań Drogowych w Warunkach Zmienności Ruchu. Monografia 347. Politechnika Krakowska, Kraków (2007)
54. Macioszek, E.: Modele Przepustowości Wlotów Skrzyżowań Typu Rondo w Warunkach Wzorcowych (Models of traffic capacity in roundabout inlets in ideal conditions). Open Access Library, vol. 3(21), Gliwice (2013)

55. Leemann, N., Santel, G: Two lane roundabouts. In: 9th Swiss Transport Research Conference. http://www.strc.ch/conferences/2009/Leemann.pdf
56. Hagring, O., Rouphail, N.M., Sorensen, H.A.: Comparison of capacity models for two lane roundabouts. Transportation Research Board Annual Meeting. http://citeseerxist.psu.edu/viewdoc/download?doi=10.1.1.965.8969&rep=rep1&type=pdf
57. Brilon, W., Bondzio, L., Weiser, F.: Experiences with turbo roundabouts in Germany. http://nmfv.dk/wp-content/uploads/2014/04/Experiences-with-Turbo-Roundabouts-in-Germany-Brilon-Bondzio-Weiser.pdf

Analysis of Selected Types of Transport Behaviour of Urban and Rural Population in the Light of Surveys

Jacek Chmielewski(✉) and Paulina Olenkowicz-Trempała

Faculty of Building, Architecture and Environmental Engineering, University of Technology and Life Sciences, Bydgoszcz, Poland
{jacek-ch, paulina.olenkowicz}@utp.edu.pl

Abstract. The article indicates that in transport analysis it is important to take account of the diversity of transport behaviour of population in the function of place of residence and area development. The results of transport behaviour of analysis have been presented according to the database of surveys from the selected Province (Kujawsko-Pomorskie, Poland). Major differences have been shown in the results of basic characteristics that describe the method of taking trips by residents of cities, towns, suburban municipalities, and rural areas. Moreover, it has been pointed out that surveys conducted in a reliable manner as well as the proper analysis of their results have the crucial importance for the correct operation of transport models, being the basic tool in verifying transport plans, thus affecting the final results of prepared documents related to transport policy in analysed areas.

Keywords: Transport plan · Transport behaviour · Surveys

1 Role of Population Surveys

One of the empirical methods of analysing transport behaviour of population are surveys, where a questionnaire is used to collect information. Surveys are diagnostic tests and involve gathering information about individual characteristics of respondents and their trips taken within a determined period.

The analysis of transport behaviour of population is a kind of quality test [1]. Surveys are usually conducted with the participation of an interviewer (a direct interview) using a questionnaire in a hard copy or on mobile devices with dedicated database software. As surveys are limited by nature, an important issue is to select a test sample correctly (a group that will be surveyed), so as to be able to generalise the results about the overall population. The most popular method of selecting the size of a test sample is a draw, taking account of the representation of specific transport areas so as to reflect optimally the demographic structure and transport behaviour of population. The size of the population needs to be taken into account. The smaller it is, the higher percentage of population should be surveyed.

Surveys into transport behaviour are usually carried out in households. Their objective is to identify the method of fulfilling transport demands. In particular,

E. Macioszek and G. Sierpiński (eds.), *Recent Advances in Traffic Engineering for Transport Networks and Systems*, Lecture Notes in Networks and Systems 21,
DOI 10.1007/978-3-319-64084-6_3

a so-called trip journal is analysed, which indicates the frequency of transport demands (mobility of residents) and how they are fulfilled. The main research issues concern trip motivations, used means of transport, as well as the time of trip and its destinations. Out of detailed research problems, the following data need to be distinguished: trip start and end times, regular trip patterns, opinions about specific transport policy solutions [2], etc. While determining transport behaviour of residents, their personal characteristics also matter, e.g. age, education, own means of transport, sex, and place of residence.

The problem of collecting data about transport behaviour, and in particular the inconvenience and costs of gathering such information, stimulate the search for newer techniques of conducting such studies. For years, research have been conducted all over the world to reduce the costs of collecting such data by aggregating study areas (so-called transport zones) [3–5] or reducing the quantity of test samples [6, 7] (a traditional approach that involves surveys at households). The application of GPS devices or tracing connections made within the mobile telephone networks are still costly and unverified techniques, with an additional problem posed by personal data protection and processing big data. However, in recent years such solutions have been applied more eagerly in studies of transport behaviour, although their functionality is limited in particular to controls of interviews conducted. [8, 9] By analogy, the use of mobile phone trace records in mobile telephony stations in order to register the trajectories of movements [10] constitutes a rather preliminary research test than the practical application of this method at present. Focus studies are a more promising method [11], where detailed interviews about transport behaviour are carried out on a small group of respondents. Such interviews are carried out by experts (usually psychologists) and recorded in order to support the process of identifying and describing transport behaviour. The technique of expert studies is also worth mentioning, where opinions of persons who are professionally involved in modelling trips are the basis for creating procedures and mathematical algorithms that describe trip processes. [12] More over data on road freight transport should be stored. Examples of how data on this traffic can be collected is shown in [13].

The outcome of surveys conducted among population within transport behaviour are databases that are the basis for developing transport models, including regional and local ones. One of the practical applications of transport models includes their use at the stage of developing transport studies or transport plans. It has to be kept in mind that a very important stage of creating models is the verification of the reliability and cohesion of data that describes both transport demands of residents and the possibilities of their fulfilment (transport infrastructure description). The verification of their correctness is a very time-consuming process, although necessary for the reliability of further analyses in the context of preparing planning documents.

The purpose of analyses carried out for this article was to verify the hypotheses that indicate the fundamental differences in transport behaviour among the population of varied urbanization levels, expressed with the density of population and the characteristics of area development. Moreover, the existence of such a dependence highlights the need to separate groups of people with homogeneous transport behaviour in the function of place of residence, in order to construct transport models.

2 Provincial Transport Plans as the Source of Data on Transport Behavior

Transport plans are strategic documents that constitute local regulations applied not only in Poland but also in other European countries as a result of implementing Regulation of the European Parliament No. 1370/2007 on public passenger transport services by rail and by road. In Germany, such plans usually comprise mid-term strategies, including: an analysis of local trip models, forecast supply and demand, basic network schemas, timetables, and funding. Some transport plans may involve a concept of developing an integrated system and different options for specific subsystems. Moreover, economic and environmental effects are estimated for individual variants. In addition, network and organizational solutions are proposed in the form of line routes, timetables and fares. Finally, transport plans describe procedures of their implementation, contracting services and monitoring as well as reporting of plan performance.

In France, on the other hand, all urban areas with population above 100 thousand are obliged to draw up city transport plans that include objectives and planned activities. For example, Chambéry in the French Alps (Savoie Department) has 59 thousand residents, while the urban area served by a local bus network has the population of 126 thousand. This is important from the point of view of the area covered by a transport plan. In France, in addition to areas attributed to specific administration units, there is also a concept of perimeter (perimeter) in transport planning that covers the area delineated due to its functions that justify the organization of integrated passenger public transport for such an area [14].

One of the main objectives of creating transport plans is to improve the conditions for the effective and safe movement of people. The obligation to prepare such plans stems explicitly from the rules of law [15]. A transport plan determines in particular: a communication network where public transport is to be provided, the assessment and forecast transport demand, anticipated funding of transport services, preferences as to the selection of the means of transport, the rules of organising the transport market, the demanded standard of public transport services, and the anticipated method of organising the IT system for passengers. Such plans should be developed at each level of rendering public transport services, from national, through provincial, to local, and municipal. In general, the implementation of such plans requires the ongoing monitoring of progress as well as their updates, in order to adapt them to current needs. Therefore, big data has to be collected continuously for such plans. Hence, data necessary for conducting analyses and monitoring the progress of implementing transport plans is the source of information, including data about transport behaviour of population as well as trends and changes in such behaviour. Consequently, it can be assumed that an additional objective of creating and monitoring the progress in the execution of transport plans is to update the databases of analysed areas in terms of demand and supply in transport, including:

- information concerning area development (enterprises, centres of education, points of commerce and services, residents, etc.),
- results of traffic volume and its characteristics,

- occupation of the means of public transport,
- results of surveys concerning transport behaviour,
- outcome of studies concerning expectations of residents in the development of transport systems,
- results of surveys into the satisfaction with public transport services, etc.

Data collected in this way makes it possible not only to adapt the transport systems to the demands and expectations of residents on the on-going basis but also enables the analysis of changes in the above-mentioned characteristics, and hence to indicate the trends of such changes. As a result, it is possible to plan and forecast better the changes of transport behaviour of population, resulting in the better parametrization of the prognostic transport models that play a key role in developing transport plans. As a result, it should contribute to the enhanced quality of transport services and the improved quality of life for residents of the area.

3 Testing Grounds

In order to verify the research hypothesis, a decision was taken to analyse results of research into transport behaviour of residents in one of 16 provinces of Poland (Kujawsko-Pomorskie) situated in the central part of the country, with population of more than 2.1 million. The surveys aimed at determining the key parameters of transport behaviour necessary to construct a transport model for the existing status were undertaken in spring 2012. 21,216 questionnaires were completed in total at 13,655 households, including 8,000 at two capitals of the Province (Bydgoszcz and Toruń), which corresponded to almost 1.5% population, while other respondents came from towns, villages, suburbia, and local municipalities (amounting to about 0.63% of the population). The distribution of the respondents was preceded with the demographic analysis of the whole Province. The number of interviews in every area was selected proportionally to the number of residents, in such as manner as to obtain a representative sample in each case (see Fig. 1). The obtained numbers fulfilled the condition of the minimum number of samples for specific characteristics in terms of mathematic statistics and the assumed level of trust in the analysis results.

The method of direct interview was used to carry out the research at households. The questionnaire was prepared in such a way as to enable the determination of selected characteristics of transport behaviour of people, necessary in the process of constructing a transport model. A questionnaire included the description of typical trips during a working day (Tuesday, Wednesday, or Thursday), which preceded directly a questionnaire day. Pollsters visited randomly selected households, where they carried out interviews with each member of the family above 9 years old, regardless of whether they took a trip or not. The reliability of collected data was verified with the internal quality control system and by persons who introduced the data on the on-going basis from questionnaire forms into the database. In addition, results of surveys carried out by KIDiT (Katedra Inżynierii Drogowej i Transportu - Uniwersytet Technologiczno-Przyrodniczy w Bydgoszczy - Department of Road and Transport Engineering, University of Science and Technology in Bydgoszcz) in 2010 were used as well.

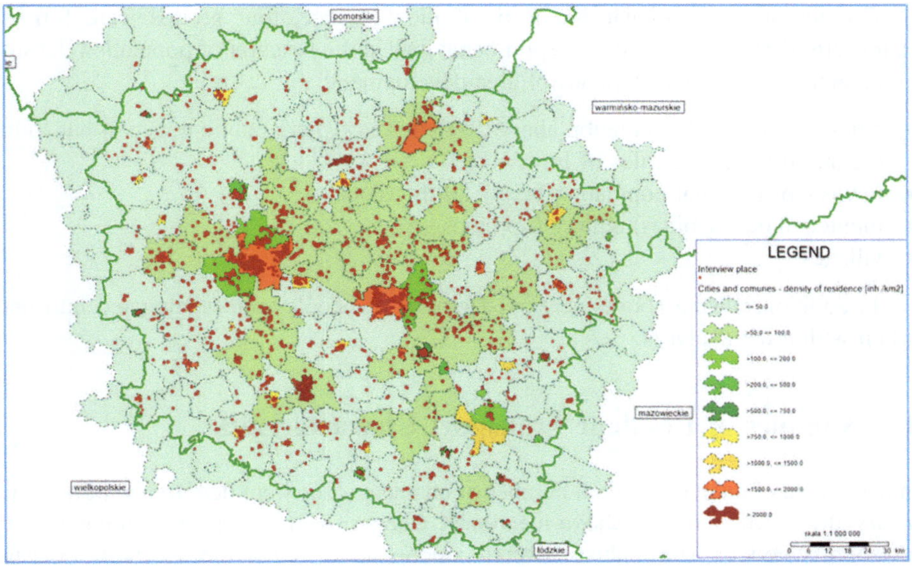

Fig. 1. Sites of questionnaires conducted among the population

At the stage of developing the characteristics of transport behaviour of the population in the Province, the following data was analysed in detail:

- trip destinations (a bus/tram stop, a station, a car park, and a garage are treated as access to means of transport in the analysis);
  - work - to a place of work and on business,
  - education - to school or university,
  - shopping - retail,
  - a shopping mall,
  - home,
  - other - private matters, recreation.
- the selection of means of transport used to take a trip:
  - on foot,
  - by car as a passenger,
  - public transport - by bus or tram, by train, etc.
  - by car as a driver,
  - by bike,
  - by tractor,
  - by motor or motorbike.
- an average distance and times of trips,
- the characteristics of the probability of selecting the means of transport to take a trip for a given distance and the trip motivation,
- the characteristics of the probability of selecting the means of transport to take a trip in the given time and the trip motivation.

Considering the variegated density of the Province, four groups were formed characterised by homogenous transport behaviour in the function of population density in the area and the characteristics of area development:

- cities - of the total population above 50 thousand (where public transport has to be organised under the rules of law),
- towns - of the total population below 50 thousand,
- municipalities (gminas) adjacent to cities,
- villages.

In addition, all the above-mentioned groups of people were aggregated into one group within the analyses.

4 Examples of Resident Transport Characteristics

In the course of analytic work, a number of characteristics were determined in order to justify the creation of the above-mentioned categories of people with homogenous transport behaviour. The analysis covered, among other, data on the selection of means of transport used to take a trip. In cities, the percentage of trips on foot and by car as a driver is highest and very similar, in both cases above 30%. In towns, the difference can be seen between the share of trips on foot at almost 50% and trips by car as a driver below 20% and 10% as a car passenger. On the other hand, in rural areas more than 40% trips are taken by car, as a driver, and more than 10% as a passenger. Residents of suburban municipalities usually travel by car as a driver as well, at almost 50%. Figure 2 shows the division of transport tasks in total trips due to the place of residence.

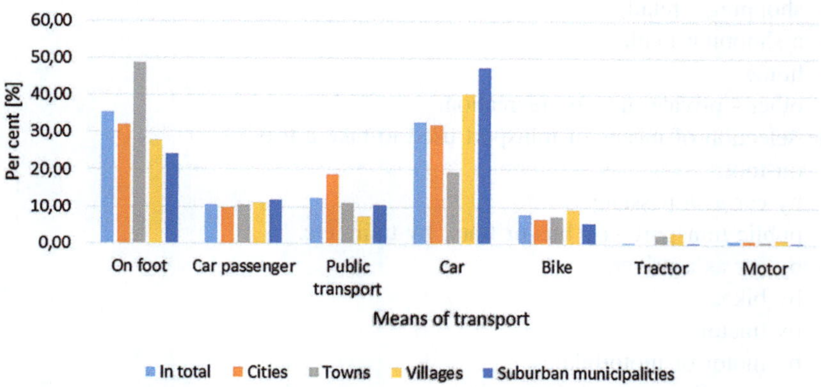

Fig. 2. Distribution of transport tasks in total trips in terms of place of residence

An important finding that results from the analysis of the transport task division is that the share of travels by individual means of transport in the largest cities is higher than by public transport. However, the difference is low. The study has also shown that

Fig. 3. Dependence between travel distance and destination due to the area of residence

residents of towns make most of their trips on foot, which is a major difference compared to trips taken by residents of cities.

Figure 3 presents the diagram describing the dependence between the distance of a trip and its destination. As shown in the diagram, the longest trips are usually done to work. An exception are trips by residents of villages and towns to shopping malls, where the distance of a trip is similar to the work destination and amounts to about 10 km. The study shows also that the shortest trips are made for retail shopping (the average distance is more than 2 km).

The analysis of the average time of trip by means of transport shows that walking is most time-consuming, while the longest trips are taken by rail by residents of cities (more than 110 min). Next longest trips are made by public transport, with an average trip time from 30 to 52 min. The differences shown in Fig. 4 between the time of trips are most evident in case of trips by public transport, for example trips by buses by city residents are shortest in the indicated group.

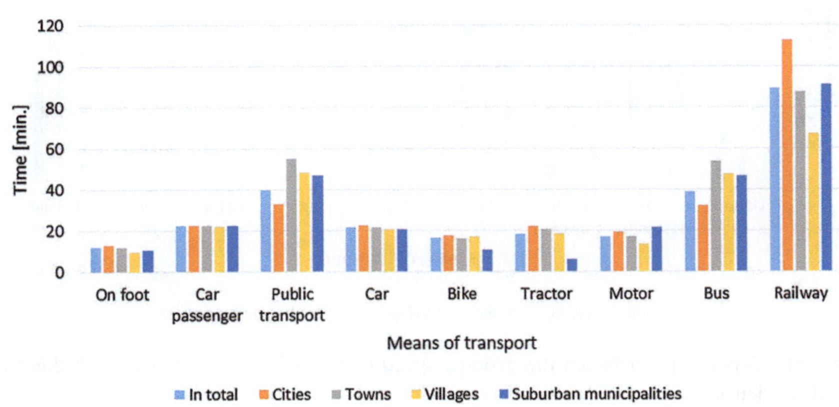

Fig. 4. The dependence between the average trip time by means of transport due to the place of residence

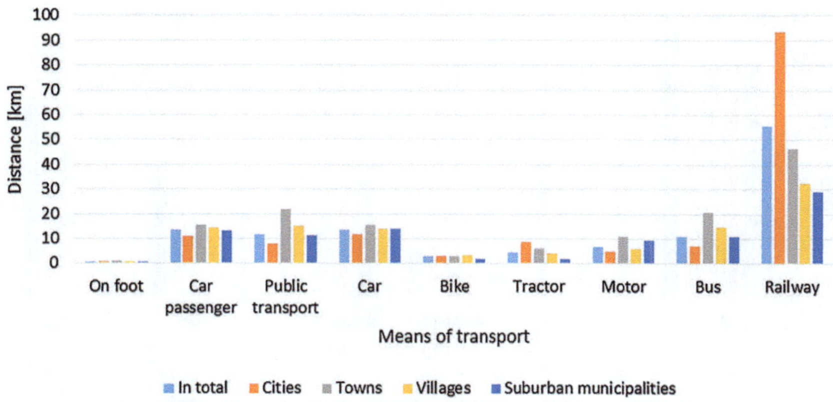

Fig. 5. The dependence between the average distance of a trip by means of transport due to the place of residence

While analysing the lengths of trips taken by the means of transport in Fig. 5, it is noteworthy that the longest trips are made by rail, with the average length of such trips for all citizens from 30 to 90 km. The shortest trips among all groups of population are on foot, which results from the general availability of this method. In public transport, the shortest trips are made by residents of cities, while the longest by people living in towns. The characteristics are similar to those in Fig. 4.

While developing the analysis results, the average speed of trips by the means of transport have been also indicated, as shown in Fig. 6. The speed by means of transport differed, with the most similar speed by bike, where an average speed was at about 12 km/h.

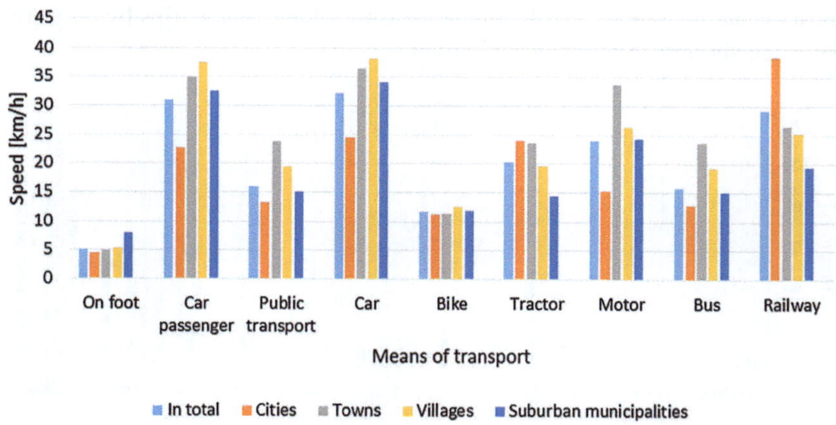

Fig. 6. The dependence between the average speed of a trip by means of transport due to the place of residence

5 Summary and Conclusions

Surveys into the transport behaviour of residents in the analysed area (province Kujawsko-Pomorskie, Poland) are of crucial importance for constructing transport models that are an essential tool in analysing transport system development, and hence in drawing up so-called transport plans. Transport plans enforce the creation and continuous updates of databases on the transport operation in an area for which they are prepared. Hence, such requirements give a chance to build bigger databases, including of transport behaviour of residents. This article shows examples of results of transport behaviour of population in the selected province and differences in the key characteristics of such trips in the function of population density and area development for the residence of respondents divided into four categories: cities, towns, villages, and suburban municipalities. Moreover, a hypothesis was confirmed that there are major differences in transport behaviour of residents in areas of different urbanization levels expressed with the density of population and the characteristics of area development. Hence, the differentiation of such characteristics shows the need to take them into account at the stage of developing transport models.

The analysed division is one of attempts at showing the variation of transport behaviour of residents in the function of place of domicile and area development. According to the authors, it is expedient to carry out further work aimed at the detailed description of the necessary division of residents into groups of homogenous transport behaviour, which are of essential importance for the detailed recreation of demand for daily trips.

References

1. Quality tests focus on characteristics of facilities, processes, and meanings that are not quantifiable (if measurable at all) in terms of quantity or size Denzin, Lincoln. http://www.wneiz.pl/nauka_wneiz/studia_inf/37-2015/si-37-55.pdf
2. Ciastoń-Ciulkin, A., Puławska, S.: Badania Zachowań Transportowych w Miastach z Wykorzystaniem Internetu. Logistyka **6**, 2750–2758 (2014)
3. Bovy, P.H.L., Jansen, G.R.M.: Network aggregation effects upon equilibrium assignment outcomes: an empirical investigation. Transp. Sci. **17**(3), 240–262 (1983)
4. Crevo, C.C.: Impacts of zonal reconfigurations on travel demand forecasts. Transp. Res. Rec. **1305**, 72–80 (1991)
5. Khatib, Z., Chang, K., Ou, Y.: Impacts of analysis zone structures on modeled statewide traffic. J. Transp. Eng. **127**(1), 31–38 (2001)
6. Smith, M.E.: Design of small-sample home-interview travel surveys. Transp. Res. Rec. **701**, 29–35 (1979)
7. Stopher, P., Kockelman, K., Greaves, S., Clifford, E.: Reducing burden and sample size in multi-day household travel survey. Transp. Res. Rec. **2064**, 12–18 (2008)
8. Giaimo, G., Anderson, R., Wargelin, L., Stopher, P.: Will it Work? Pilot results from the first large-scale global positioning system-based household survey in the United States. J. Transp. Res. Board **2176**, 26–34 (2010)

9. Swann, N., Stopher, P.: Evaluation of a GPS survey by means of focus groups. Transportation Research Board, Washington (2008)
10. Friedrich, M., Immisch, K., Jehlicka, P., Otterstatter, T., Schlaich, J.: Generating origin-destination matrices from mobile phone trajectories. J. Transp. Res. Board **2196**, 93–101 (2010)
11. Dziedzic, T., Szarata, A.: Wybrane Wyniki Kompleksowych Badań Ruchu w Województwie Małopolskim. In: IX Poznańska Międzynarodowa Konferencja Naukowo-Techniczna z cyklu Problemy Komunikacyjne Miast w Warunkach Zatłoczenia Motoryzacyjnego, pp. 19–21. SITK RP oddział w Poznaniu, Poznań (2013)
12. Szarata, A.: Kompleksowe Badania Ruchu w Krakowie w 2013 r.—Wybrane Wyniki. Logistyka, vol. 6 (2014)
13. Kijewska, K., Małecki, K., Iwan, S.: Analysis of data for urban transport systems. In: Mikulski, J. (ed.) Challenge of Transport Telematics. CCIS, vol. 640, pp. 135–148. Springer, Switzerland (2016)
14. Website. http://www.rbf.net.pl/wiecej,Czy-plany-transportowe-sa-potrzebne,736
15. Law Office of the Parliament: Act of 16 December 2010 on Public Transport - Journal of Laws from 2011, No. 5, Item 13 and Regulation by the Minister of Infrastructure of 25 May 2011 on the Detailed Scope of the Plan of Public Transport Sustainable Development—Journal of Laws from 2011, No. 117, Item 684. Law Office of the Parliament, Warsaw (2011)

Selected Aspects of the Methodology of Traffic Flows Surveys and Measurements on an Urban Agglomeration Scale with Regard to ITS Projects

Renata Żochowska[1(✉)], Grzegorz Karoń[1], Ryszard Janecki[2], and Aleksander Sobota[1]

[1] Faculty of Transport, Silesian University of Technology, Katowice, Poland
{renata.zochowska, grzegorz.karon, aleksander.sobota}@polsl.pl

[2] Faculty of Economics, University of Economics, Katowice, Poland
ryszard.janecki@ue.katowice.pl

Abstract. The article presents the scope and the algorithm of traffic flows surveys and measurements (TFSM) necessary for the construction of an urban agglomeration transportation model (UATM) for ITS projects. The basic guidelines for determining the locations of the measurement points have also been discussed. To show the structure of TFSM the algorithm of action during Comprehensive Traffic and Transportation Studies for Transportation Modeling with Evaluation of ITS Project, has been presented. The examples relate in particular to urban agglomeration with the nature of a conurbation - Upper Silesian Agglomeration in Poland.

Keywords: Comprehensive traffic and transportation study · Measurement points · Traffic flows surveys and measurements · Agglomeration transportation model

1 Introduction

Traffic flows surveys and measurements (TFSMs) [1, 2] constitute one of the essential stages of construction of the urban agglomeration transportation model (UATM) [3–5]. Therefore, an important issue is to design activities in this phase, including:

- establishing the basic scope of TFSMs necessary for the construction of the transportation model,
- determination of the additional scope of TFSMs that results from the characteristic of the transportation project, e.g. from the scope of ITS project [6, 7],
- identification of existing data and assessment of their usability in the construction of UATM - data that may be used to reduce the scope of TFSNs or to supplement them.

A review of the state-of-the-art revealed that issues that need to be treated comprehensively in methodology of traffic flows surveys and measurements on an urban

E. Macioszek and G. Sierpiński (eds.), *Recent Advances in Traffic Engineering for Transport Networks and Systems*, Lecture Notes in Networks and Systems 21,
DOI 10.1007/978-3-319-64084-6_4

agglomeration scale with regard to ITS projects [8] are described in foreign literature well, but too general [1–5]. In turn, there are many publications in the national literature that deal with these issues in a fragmentary or superficially manner. However, there is no comprehensive work describing in a methodical.

Many authors of various publications describe insufficiently the elements of the developed methodology, which involve such questions as:

- which components of ITS are most important in comprehensive traffic and transportation studies for transportation modeling for evaluation of ITS project,
- how to include structure of ITS project above all: stakeholder aspirations, functional and physical architecture, ITS subsystems, and ITS user services in survey of transportation behavior and preferences,

Developed methodology with algorithm of traffic flows surveys and measurements on an urban agglomeration in transportation models for the evaluation of ITS projects - which uses a systematic approach and a formalization of the key research questions - is an important scientific contribution.

2 Scope of TFSMs Necessary to Construct the Urban Agglomeration Transportation Model

Developing the transportation model [9–12] requires gathering data on trips and on traffic flows [13, 14], taking into account the following characteristics [1, 2]:

- categories of trips and traffic flows:

– internal trips and traffic flows - generated by the residents and vehicles stationed in the study area,
– external trips and traffic flows - generated by persons and vehicles from outside the study area entering the area for less than one day.

- trips and traffic flows divided by their origin and destination (see Fig. 1, item 0):

– internal trips and traffic flows - generated by the residents and vehicles stationed in the study area,
– internal-external trips and traffic flows - started in the study area and ended outside the area,
– external-internal trips and traffic flows - started outside the boundary of the study area and ended in the area,
– crossing trips and traffic flows - started and ended outside the study area but passing through the area,
– external trips and traffic flows - started and ended outside the study area and not passing through the area, but passing through its neighborhood; knowledge of external-only traffic flows (trips) may be important for transportation analysis of investment projects that significantly affect the neighborhood of the study area - by increasing transportation accessibility between the area and its surroundings or by investing in commercial and service activities, which would change the part of the

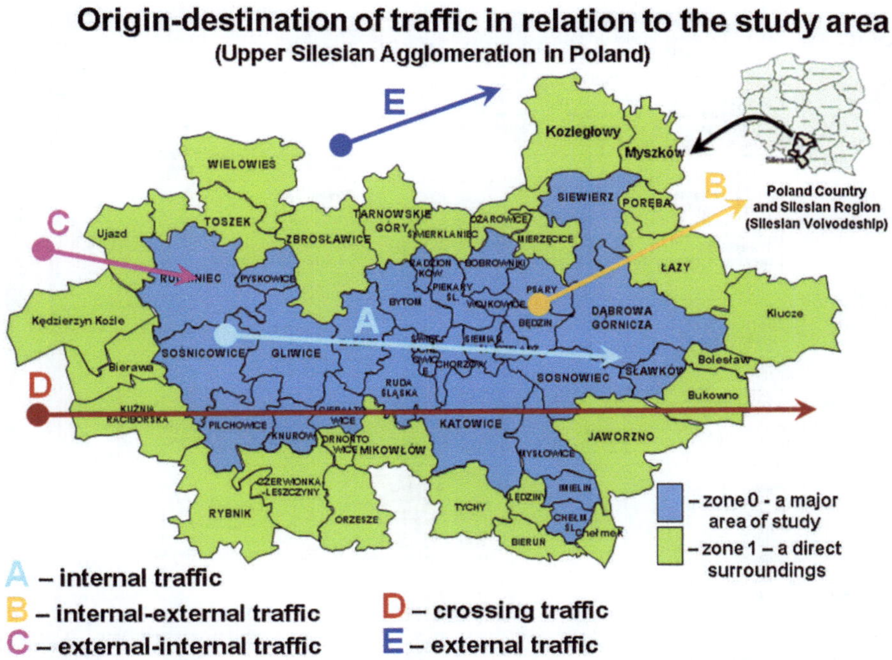

Fig. 1. Traffic in relation to the study area (Source own study based on [8])

current external-only traffic flows (trips) into destination traffic flows (trips) or into passing through traffic flows (trips).

Traffic in relation to the study area has been presented on the example of the area of Upper Silesian Agglomeration (in Poland), in which the assumptions of discussed methodology has been verified (see Fig. 1).

The presented trips and traffic flows characteristics require comprehensive approach to carry out TFSMs. This means performing them relatively simultaneously at specific measurement points and for a relatively short period of time. The term "relative" refers - in this case - to the real (practical) conditions for the implementation of comprehensive traffic and transportation studies (CTTSs), which relate to the technical and organizational possibilities of the CTTSs contractor as well as to the financial possibilities of the CTTSs contracting entity. To show the structure of TFSM in CTTM the algorithm of action during Comprehensive Traffic and Transportation Studies for Transportation Modeling with Evaluation of ITS Project, has been presented in Fig. 2. As part of the CTTSs it covers the following types of studies:

- questionnaire surveys (see Fig. 2 block 4 and next) carried out on random samples of persons and vehicles collected during:

– household interviews,
– interviews in transportation bases - among cargo carriers,

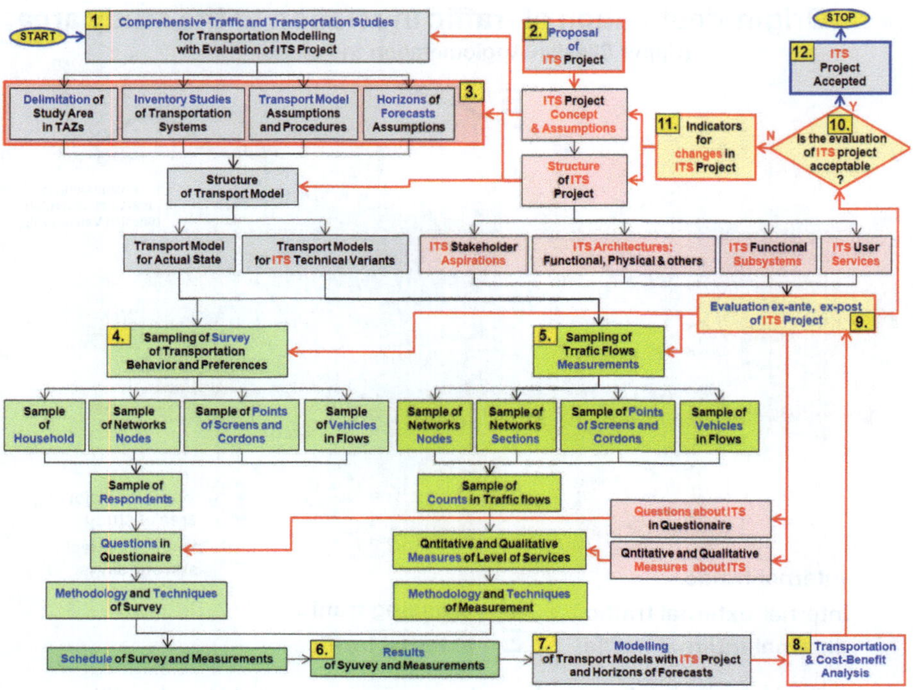

Fig. 2. The algorithm of action during Comprehensive Traffic and Transportation Studies (CTTS) for Transportation Modeling with Evaluation of ITS Project (in form of block-scheme)

– interviews with users of specific vehicles - e.g. taxis, car sharing, autonomous vehicles [15],
– interviews with car drivers - at cordon points, car parks,
– interviews with passengers of public transport - at cordon points, stations, stops and multimodal integrated nodes,
– interviews with pedestrians and cyclists - at the points of traffic concentration of these users.

The results of the study should make it possible to identify trip origins and destinations as well as to identify transportation characteristics of:

• persons - among others, travel behavior and transportation preferences [16], including the assessment of satisfaction with specific technical solutions of the ITS project [6, 7, 17], the usability of these solutions, and the impact of this usability on the travel behavior of transportation users,
• vehicles - including mobility index, origin and destination of the ride,
• spatial area - among others, type of activity and trip purpose for the objects generating destination traffic flows (trips),
• inventory studies in the following scopes (see Fig. 2 block 3):

- functional and organizational network identification of transportation infrastructure facilities: (a) network function of the following facilities: road and street sections, intersections, road junctions, car parks and parking lots, stops, stations and multimodal/integrated nodes, cycling routes, station of bicycle rental systems, elements of pedestrian infrastructure; (b) system organization - such as the type of organization and traffic control on the sections, intersections and road junctions, capacity of car parks and parking spaces including the toll collection system and authorization to entry to the stopping place, the routes of public transport lines with timetables, city bike rentals along with the toll collection and bike lending system, car sharing with toll collection and car lending system,
- identification of the object of land-use development generating destination traffic flows (trips) - functional characteristics of the objects in terms of: (a) type of activity, hours of operation, user groups, trip purpose and potential mobility; (b) passive transportation availability (possibility of a journey to a particular destination from other source sites) and active transportation availability (possibility of a journey from specific source sites to other destinations), usability of transportation systems serving the object - including space and time availability of parking spaces, stations and stops, and public transport lines serving the stations and stops in the vicinity of the object,
- identification and assessment of usefulness of available data on traffic flows and transportation models of the internal zones of the study area (municipalities and associations of municipalities as the components of the study area) and the zones surrounding the study area (subregion, region/voivodeship, country), concerning among others: (a) traffic flow data - identification and assessment of the results of traffic flow measurements, the results of passenger flow measurements in public transport (including rail transport), the results of questionnaire surveys on travel behavior and transportation preferences; the source of these data may be data collected during construction of transportation models, results of General Traffic Measurements (GPR 2015, 2010, 2005), data collected while carrying out the projects of technical and organizational modernization of intersections, road junctions and other objects of transportation infrastructure; (b) data on transportation models - identification and assessment of models in the following areas: number and the characteristics of traffic analysis zones (e.g. number of inhabitants and their age structure, number of students, numbers of jobs, etc.), representation of the transportation network (determination of parameters of the nodes as well as of the road and street network segments, detailed mapping of public transport networks, mapping of pedestrian and bicycle infrastructure elements), types of trip purpose, groups of users with homogeneous travel behaviors, methods of mapping of external, origin, destination and passing through traffic flows, methods of mapping of travel behavior and transportation preferences, and division of transportation tasks (e.g. utility functions and the choice models),

- traffic flow measurements (see Fig. 2 block 5 and next):

– flows of persons/passengers in vehicles, at stops and stations of public transport, at car parks (including carshare system), at the stations of bike rental systems, at selected pedestrian routes,
– flows of people entering and leaving the certain objects - the mobility generated by specific objects,
– flows of vehicles at specific points of the transportation network (approaches of the intersections and road junctions, selected points on road and streets sections, parking entrances with parking measurements, i.e. parking lot rotation indicator, parking time, etc., divided into groups of vehicles and directional structure,

- studies and measurements of traffic conditions (see Fig. 2 emphasized block below block 5):

– measurements of travel times and delays at the approaches of the intersections (number of stops and queue lengths) on main street lines - for individual vehicles as well as for public transport vehicles (additionally punctuality measurements),
– measurements of the speed of passing through the sections of main transportation corridors, depending on their traffic load; the goal is to determine the so-called link impedance function needed to calibrate the transportation network,

- studies and measurements aimed at ex-ante and ex-post evaluation of the results of a specific investment project using the transportation model such as the ITS project [6, 7, 11] (see Fig. 2 block 9 and next):

– surveys of the assessment of user satisfaction with specific technical solutions of the ITS project, the usability of these solutions, and the impact of this usability on the travel behavior and transportation preferences of the users,
– measurements of traffic conditions adapted to the purpose of the ITS project and its impact - spatial and temporal extent and impact on the usability of transportation systems as well as on the travel behavior of the users and their transportation preferences.

3 Guidelines for Determining the Points of TFSMs

3.1 Measurement Points on the Screens and Cordons of the Study Area

Measurements of traffic flows at the screen and cordon points (see Fig. 2 emphasized blocks below blocks 4 and 5) are carried out at locations where links of the transportation network cross the established boundaries of the study area. These are screen points - determined by the crossing at open lines and cordon points - determined by the crossing at closed lines. The basic measuring range is [16]:

- measurement of traffic flows of vehicles and/or persons (occupancy of the vehicles) along with the structure of vehicles,

- questionnaire surveys of people at selected points,
- measurement of local/instantaneous velocity - as additional measurements related to the investment project, e.g. to calibrate the impedance function of the section.

Criteria for determining measurement points on the screens and cordons [16]:

- type of area delimitation:

– linear (open) delimitation - lines of the screen, such as lines along rivers, railways, motorways and other established open lines separating parts of the area,
– circular delimited (closed) - cordon lines e.g. the boundary of the agglomeration, the boundary of the city center, downtown, etc.,

- location of the screen and cordon points:

– determined by the crossing the screen and cordon lines by all the major roads and lines of public transport,
– recommended a limited number of the screen and cordon points - to those where there is a significant volume of traffic flow between areas divided by screen or cordon lines,
– clearness of the area division - the screen or cordon lines should follow the administrative boundary of the agglomeration and along the administrative boundaries of the towns and cities of the agglomeration.

3.2 Measurement Points in the Transportation Network Within the Study Area

The type, location and number of measurement points in the transportation network (see Fig. 2 emphasized blocks below blocks 4 and 5) within the study area are primarily related to the following objects of the transportation infrastructure [16]:

- intersections, road junctions and sections between these objects,
- stops and stations of public transport,
- multimodal transportation nodes with integrated parking systems (Park & Ride, Bike & Ride, etc.),
- car parks with a large surface area in the surroundings of shopping and service centers, other car parks and parking spaces constituting a concentration of origin and destination traffic, parking areas in paid/authorized parking areas - especially for the ITS projects related to the strategies of transport management utilizing the management of access to parking spaces and guidance to free parking spaces using, for example, VMS tables,
- bicycle routes, stations of the city bike rentals and elements of pedestrian infrastructure.

The choice of the individual points of the mentioned objects of the transportation infrastructure requires an appropriate analysis, in which the decisive criteria for the selection include [16]:

- significance of the results of TFSMs in transportation model:
  - in base variant of the model, describing the current situation,
  - in the technical and forecasting variants of the model, describing the ITS project in certain variants (configurations) and assessing the results of these variants within predetermined horizons,
- scope of the surveys and scope of measurements:
  - some surveys, such as questionnaire surveys of the users on their travel behavior and transportation preferences, may be realized among the passengers stationary, i.e. at the stops, while waiting for the vehicle; other surveys may or must be carried out in motion, i.e. in vehicles - e.g. they must be performed in vehicles exceeding the screen or cordon points, because it is not possible then to perform these surveys stationary,
  - some measurements are stationary (e.g. passenger counting, vehicle counting, queue measurements at intersections, local/instantaneous velocity measurements) and can be made at stops, intersections, parking lots; other measurements, such as measuring the average speed of vehicles in free flow and in various states of traffic saturation of road and street sections - measurements for the calibration of the impedance function of the sections - may be conducted using test vehicles in combination with the stationary measurements of traffic volumes or density on the analyzed sections,
- methods and techniques (see Fig. 2 emphasized blocks below blocks 4 and 5) for carrying out TFSMs - both in the surveys of travel behavior and transportation preferences and in the measurement of traffic flow characteristics, various techniques and research methods are used, as follows:
  - stationary - at measurement points with the help of interviewers, observers and devices temporarily installed for testing purposes,
  - on the move - in moving public transport vehicles or using test vehicles,
  - remote or off-line - internet surveys, mobile data applications for travel data, data from data repositories in traffic control and traffic management centers (ITS), data from the continuous traffic data systems, data from automated passenger and vehicle counting systems (counting gateways, systems of the electronic ticket for the ride, parking fee systems, toll collection systems on toll roads),
- the expected volume of traffic flows: vehicles, passengers:
  - the estimated/expected volume of the flows of passengers and vehicles is necessary to determine the size of the test sample, the time interval at which the sample will be tested, and the schedule of the individual TFSMs for a particular study period,
  - the accepted value of the test sample, for particular types of TFSMs, determines in large part the selection of measurement points in terms of the expected intensity of recorded data and of the intensity of interviews,

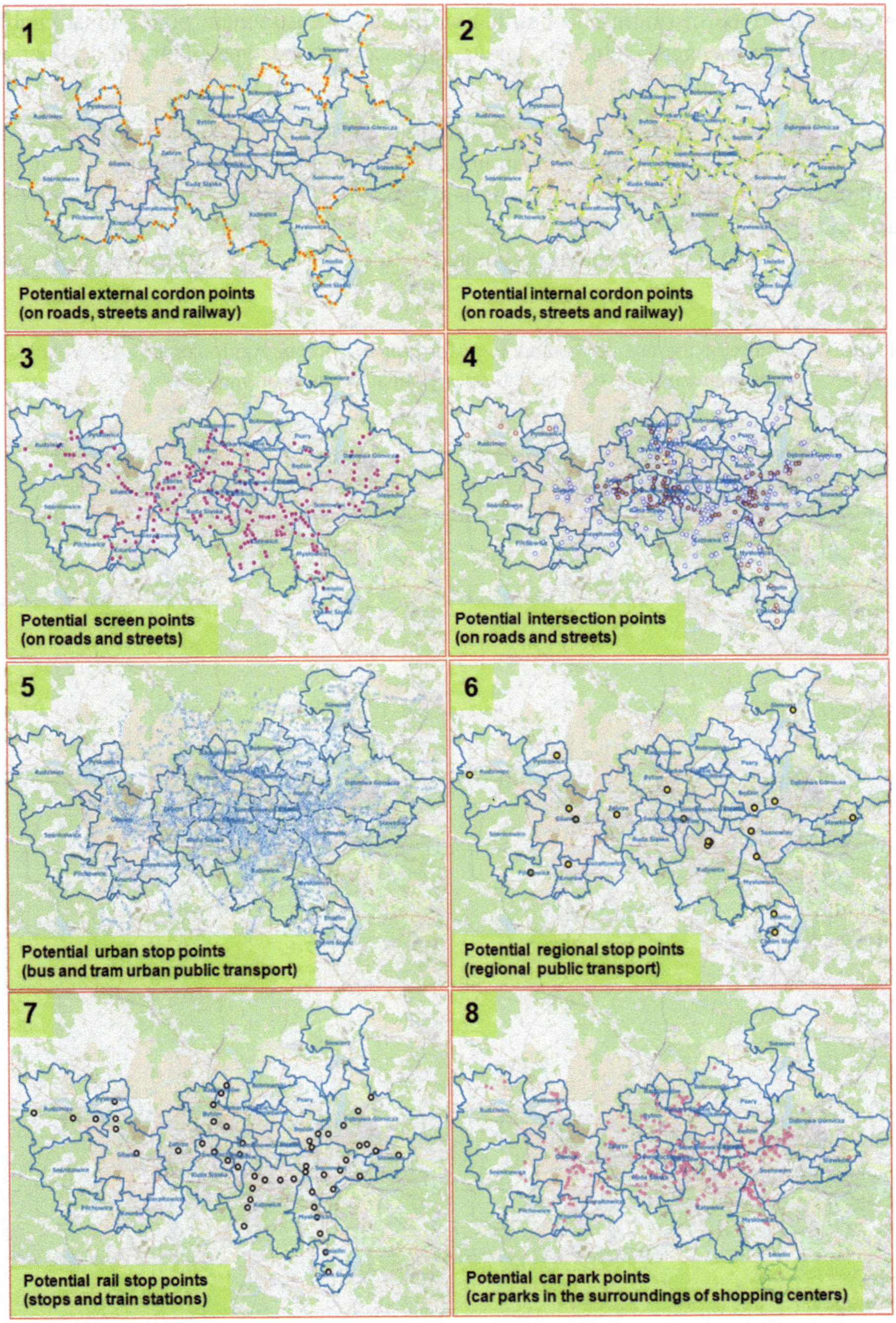

Fig. 3. Results of using the methodology for Upper-Silesian Agglomeration - the potential measurement points: 1, 2 - external and internal cordon points, 3 - the screen points, 4 - intersections, 5, 6, 7 - bus, tram and train stops, 8 - car parks

- expected type of origin and destination traffic - measurement points must enable TFSMs (passengers, vehicles), taking into account the type of origin-destination relationships, relative to the study area (see Fig. 1, item 0).

The developed methodology and algorithm presented in Fig. 2 allow to prepare the transportation models and the evaluation of ITS projects on different scales, i.e. on the scale of one of the interior cities of the agglomeration (e.g. for the city of Katowice according to the ITS concept [7]), on the scale of the whole agglomeration (e.g. for the Upper Silesian Agglomeration as elaborated by the ITS concept [6]) as well as for the

Table 1. Results of the use of methodology for Upper Silesian Agglomeration (in Poland) – number of measurement point of selected types estimated based on methodology.

Administrative criterion of delimitation – cities of urban agglomeration	Type measurement points		
	Number of intersections with bus lines	Number of intersections with bus and tram lines	Number of intersections with tram lines
Będzin	5	2	0
Bobrowniki	3	0	0
Bytom	9	13	2
Chełm Śląski	0	0	0
Chorzów	3	7	1
Czeladź	3	0	0
Dąbrowa Górnicza	9	5	0
Gierałtowice	1	0	0
Gliwice	8	0	0
Imielin	0	0	0
Katowice	20	4	0
Knurów	4	0	0
Mysłowice	4	0	0
Piekary Śląskie	2	0	0
Pilchowice	3	0	0
Psary	2	0	0
Pyskowice	1	0	0
Ruda Śląska	2	4	0
Rudziniec	1	0	0
Siemianowice Śląskie	4	1	0
Siewierz	1	0	0
Sławków	1	0	0
Sosnowiec	5	14	0
Sośnicowice	1	0	0
Świętochłowice	0	1	0
Zabrze	3	14	0
Sum	95	65	3
Total	163		

metropolitan area of this conurbation - the area of the Central Subregion of the Silesian Voivodeship. Verification of the assumptions of this methodology was made during the preparation of two tender documentations (among others Specification of the essential terms of the contract and Description of the subject of the order): for the Upper Silesian Agglomeration and for the metropolitan area of this conurbation - the Central Subregion of the Silesian Voivodeship. One of the many results of using the methodology for Upper-Silesian Agglomeration are approximate locations of measurement points for this urban area - see Fig. 3. However, it should be kept in mind that these are only potential measurement points, because the developed methodology provides guidelines that the contractor of TFSMs should use when determining the exact number and locations of these points. Results of the use of methodology for Upper Silesian Agglomeration (in Poland) - number of measurement point of selected types estimated based on methodology have been shown in Table 1.

4 Formal Description of the Issue

The development of the methodology with the algorithm (Fig. 2) requires a formal description of the problem for the algorithmic process. This process will allow the transformation of the issues formulated in the form of a verbal description into the algorithm and then into the specialized software - decision support tool when planning activities shown in Fig. 2.

The mathematical modeling methods have been used to describe the formal way of measuring traffic flows. Consequently, the set of types of TFSMs has been defined as:

$$I = \{1,\ldots,i,\ldots,\bar{I}\} \tag{1}$$

where:

i - number corresponding to the type of TFSMs,

$\bar{I}$ - number of TFSMs types (size of the set I).

For each type of study, the measurement points have been determined. Therefore, the set containing the numbers of the measurement points for the i-th type of TFSMs has been written as:

$$\mathrm{MP}_i = \left\{ mp_i \quad : \; mp_i = 1,\ldots,\overline{MP_i}\right\}, \qquad i \in \mathrm{I} \tag{2}$$

where:

mp_i - number of measurement point for i-th type of TFSMs,

$\overline{MP_i}$ - number of all measurement points for i-th type of TFSMs (size of the set $\mathrm{MP_i}$).

For each i-th type of study, the characteristics describing the manner and conditions of carrying out this study have been determined. These characteristics may be formulated in functions of varying complexity, depending on the needs and purpose of the research. Thus, they may be expressed in the form of specific values (e.g. location of the measurement site, number of people assigned to conduct research in a particular place, minimum sample size, etc.), in the form of a set (e.g. a list of measured values in the measurement site, number of questions in a survey, etc.) or as a complex procedure (e.g. description of measurement technique, measurement schedule, etc.).

The number of characteristics for each type of the study may also vary. Therefore, a set J_i containing numbers of characteristic for *i*-th type of TFSMs has been defined as:

$$J_i = \{j_i \quad : j_i = 1, \ldots, \overline{J_i}\}, \qquad i \in I \tag{3}$$

where:

j_i - number of function characterizing the *i*-th type of TFSMs,

$\overline{J_i}$ - number of all functions characterizing the *i*-th type of TFSMs (size of the set J_i).

Therefore, the functions (characteristics) used to describe the *i*-th type of study have been described in form of the vector:

$$FMP_i = \langle fmp_{j_i,i} : j_i \in J_i \rangle, \; i \in I \tag{4}$$

where:

$fmp_{j_i,i}$ - vector containing the values of j_i-th characteristic for each measurement point when performing the *i*-th type of study.

5 Conclusions

- The issues presented in the article, concerning the scope of TFSMs with their formal description, in the case of preparation of the description of the contract subject for the contractor, should be supplemented with the following tasks:
- developing a detailed schedule for conducting TFSMs,
- obtaining the necessary permits and agreements for the implementation of TFSMs,
- developing the information and promotion actions directed to the users of transportation system and residents of the study area,
- preparing the description of the contract subject for the contractor of TFSMs.

Further research will be focused on the development of the specialized decision support software when planning activities shown in algorithm described in this article. It is particularly advantageous in this undertaking to use the experience of verification of the assumptions of described methodology during preparation of two tender documentations for the Upper Silesian Conurbation and for the metropolitan area of this conurbation - the Central Subregion of the Silesian Voivodeship - and experience based on projects of ITS [6, 7, 16].

References

1. Ortuzar, J.D., Willumsen, L.G.: Modelling Transport, 3rd edn. Wiley, Chichester (2001)
2. Cascetta, E.: Transportation Systems Analysis. Models and Applications. 2nd edn. Springer Optimization and Its Application, New York (2009)

3. Boyce, D., Williams, H.: Forecasting Urban Travel. Past, Present and Future. Edward Elgar Publishing, Cheltenham (2015)
4. Cambridge Systematic: Travel Survey Manual. Cambridge Systematic, Cambridge (1994)
5. Richardson, A.J., Ampt, E.S., Meyburg, A.H.: Survey Methods for Transport Planning. Eucalyptus Press, Melbourne (1995)
6. Karoń, G., Mikulski, J.: Expertise of The Concept and Architecture of the Intelligent Traffic Management System (ITS) for the Territory of the Public Transport Operator in Upper Silesian Industrial Region. Report of Research Work for Municipal Transport Union of the Upper Silesian Industrial District (KZK GOP). Katowice, Poland (2015)
7. Karoń, G., Mikulski, J.: Expertise of The Functional Analysis for Intelligent Transportation Management System (ITS) for Katowice City. Report of Research Work for City Hall of Katowice, Katowice (2011)
8. Karoń, G., Żochowska, R., Sobota, A. et al.: The Methodology and Detailed Concept of Traffic Surveys and Measurements and Transportation Model Development for the Territory of the Public Transport Operator in Upper Silesian Industrial Region (in Poland). Report of Research Work for Municipal Transport Union of the Upper Silesian Industrial District. Silesian University of Technology, Katowice (2016)
9. Jacyna, M., Wasiak, M., Lewczuk, K., Kłodawski, M.: Simulation model of transport system of Poland as tool for developing sustainable transport. Arch. Transport **31**, 23–35 (2014)
10. Jacyna-Gołda, I., Żak, J., Gołębiowski, P.: Models of traffic flow distribution for various scenarios of the development of proecological transport system. Arch. Transp. **32**, 17–28 (2014)
11. Pyza, D.: Multi-criteria evaluation of transportation systems in supply chains. Arch. Transp. **23**, 47–65 (2011)
12. Karoń, G.: Travel demand and transportation supply modeling for agglomeration without transportation model. In: Mikulski, J. (ed.) Activities of Transport Telematics. CCIS, vol. 395, pp. 284–293. Springer, Heidelberg (2013)
13. Celiński, I., Krawiec, S., Macioszek, E., Sierpiński, G.: The analysis of travellers behaviour in the upper silesian conurbation. Arch. Transp. **24**, 441–461 (2012)
14. Chamier-Gliszczyński, N.: City logistics—sustainable urban mobility. In: Carpathian Logistics Congress. CLC 2015. Conference Proceedings, pp. 263–268. Tanger Ltd., Ostrava (2016)
15. Mahmassani, H.S., Talebpour, A.: Influence of connected and autonomous vehicles on traffic flow stability and throughput. Transp. Res. Part C **71**, 143–163 (2016)
16. Kruszyna, M., Rychlewski, J.: Influence of approaching tram on behaviour of pedestrians in signalised crosswalks in Poland. Accid. Anal. Prev. **55**, 185–191 (2013)
17. Kucharski, R., Gentile, G.: Modelling information spread processes in dynamic traffic networks. In: Mikulski, J. (ed.) Challenge of Transport Telematics. CCIS, vol. 640, pp. 317–328. Springer, Switzerland (2016)

The Warehouse Location Problem in the Context of Vehicle Routing Problem in the Production Companies

Ilona Jacyna-Gołda[1(✉)] and Mariusz Izdebski[2]

[1] Faculty of Production Engineering, Warsaw University of Technology, Warsaw, Poland
i.jacyna-golda@wip.pw.edu.pl
[2] Faculty of Transport, Warsaw University of Technology, Warsaw, Poland
mizdeb@wt.pw.edu.pl

Abstract. This paper presents warehouse location-routing problem in the logistics network. In order to solve this problem the location model was developed. The limitations and the optimization criterion of the model were determined. The optimization criterion refers to the distance between the facilities of the network. The final location of warehouse facilities was obtained using an ant algorithm. This paper describes the stages of the ant algorithm i.e. the stage of designating the probability of transition of ants to the other points of the route, update pheromone. In this paper the process of calibration of this algorithm was presented. The results of the ant algorithm were compared with the random results.

Keywords: Ant algorithm · Optimization · Warehouse location-routing problem

1 Introduction

The classical issue of warehouse location is defined in literature as the capacitated warehouse location problem CWLP [1]. In this class of problems the structure of the transport and logistics network consists of warehouse facilities and customers. The main objective in the general warehouses location problem is to find such a location of warehouse facilities for which the costs arising from forwarding a specific volume of goods to the customers via warehouses are as low as possible. If no capacity limitation is imposed, an incapacitated warehouse location problem is considered [2].

The problem of warehouse situation may be generally reviewed from two basic viewpoints. The first one concerns selecting the place where a new warehouse is to be erected [1, 3]. Basically in such a case potential locations for the construction of a new warehouse are determined. The second aspect includes a review of the selection of warehouse locations from the already existing facilities, which points to a reconfiguration of the logistic network for various types of enterprises [4].

The warehouse location problem in the logistics network is multi-criteria optimization problem [5–9] which depends on quantitative and qualitative criteria [10, 11].

E. Macioszek and G. Sierpiński (eds.), *Recent Advances in Traffic Engineering for Transport Networks and Systems*, Lecture Notes in Networks and Systems 21,
DOI 10.1007/978-3-319-64084-6_5

The problem of locating warehouse facilities is also studied in the context of planning the routes that vehicles have to cover. In this case it is defined in literature as warehouse location-routing problem [12, 13]. The objective is to determine the location of warehouse facilities and goods delivery routes from warehouses to each customer, so that the cost arising from transport of a defined volume of goods to the customers is the lowest possible. This takes into consideration a situation in which the route of each vehicle contains a few unloading points (customers). The warehouse location-routing problem presented in this paper refers to the location in the context of a reconfiguration of the logistic network.

The main aim of this article is to develop the algorithm which may be used in the warehouse location-routing problem in the logistic network. In order to solve this problem the ant algorithm was developed. The application of the ant algorithm in the warehouse location-routing problem was dictated by the fact that the authors do not find the application of this algorithm in the examined problem.

Ant algorithms belong to a group of heuristic algorithms, which do not guarantee the optimal solution, but only close to the optimal solution so-called sub-optimal. Despite this inconvenience ant algorithms are a practical tool for optimization and are used in a variety of complex decision problems e.g. vehicles routing problems [14–16]. The ant algorithm solves the warehouse location-routing problem in the logistics network of the production companies. The complexity of the problem of locating warehouse facilities limits the use of accurate methods to find optimal solutions and allows to accept sub-optimal solutions. Fast time of generating the result by this algorithm is its main feature, what is desired by the companies to designate the transport routes. These routes depend on the quantity of cargo which is ordered by the companies. This quantity depends on many factors e.g. processing capacity of companies, the production capacities of suppliers. The algorithm for designating warehouse location-routing problem must be adapted to frequent changes of these factors and generate the solutions in a quickly way. In the production companies the time of generating the solution by the algorithm plays the most important role.

2 The Mathematical Model of the Problem

2.1 Assumptions

The adopted assumptions of the mathematical model refer to the product companies. These assumptions result from the general rules which functions in product companies [4]:

- transport between suppliers and the warehouse is on the side of the production company. The company collects the raw material from the suppliers and transports it to the warehouse. From the warehouse the vehicle returns to the production companies and ends the route or starts the new route from that warehouse. The return to the production company depends on the range of time in which the collect must be realized,
- the volume of the raw material offered by the suppliers meets the demand of production companies,

- the collect of raw materials from suppliers and transport to the warehouses takes place in set period of time e.g. two days,
- in the case when the volume of the raw material from the deliver is bigger than current capacity of the vehicle, the vehicle is loaded to the full and goes to the warehouse,
- the raw material for each company may be stored in various warehouses,
- transport of the raw material from the warehouses to the companies takes place after transporting the whole raw material from suppliers to the warehouses,
- the vehicle leaves the company once and returns to the company once,
- the starting point of each route is the production company or the warehouse, the ending point is always the warehouse.

2.2 The Input Data

The following variables have been formulated for the needs of establishing a warehouse location problem:

- $\boldsymbol{V} = \{v\text{: } v = 1,2, \ldots, v', \ldots, V\}$ - set of numbers of spot elements of the logistical network: suppliers, warehouses, production companies,
- $\boldsymbol{DS} = \{v\text{: } \alpha(v) = 0 \text{ for } v \in \boldsymbol{V}\}$ - set of numbers of suppliers, $\alpha(v) = \{0,1,2\}$ - the mapping which assigns the number of elements of the logistic network to object type,
- $\boldsymbol{MS} = \{v\text{: } \alpha(v) = 1 \text{ for } v \in \boldsymbol{V}\}$ - set of numbers of warehouses,
- $\boldsymbol{P} = \{v\text{: } \alpha(v) = 2 \text{ for } v \in \boldsymbol{V}\}$ - set of numbers of production companies (recipients),
- $\boldsymbol{VEH} = \{1, \ldots, veh, \ldots, VEH\}$ - set of numbers of vehicles,
- $\boldsymbol{K} = \{1, \ldots, k, \ldots, K\}$ - set of numbers of all routs,
- $\mathbf{D1} = [d1(v, v')\text{: } d1(v,v') \in \boldsymbol{R}^{+}, v \in \boldsymbol{P}, v' \in \boldsymbol{DS}]$ - distance matrix in relations: production companies - suppliers,
- $\mathbf{D2} = [d2(v,v')\text{: } d2(v,v') \in \boldsymbol{R}^{+}, v \in \boldsymbol{DS}, v' \in \boldsymbol{DS}]$ - distance: suppliers - suppliers,
- $\mathbf{D3} = [d3(v,v')\text{: } d3(v,v') \in \boldsymbol{R}^{+}, v \in \boldsymbol{DS}, v' \in \boldsymbol{MS}]$ - distance: suppliers - warehouses,
- $\mathbf{D4} = [d4(v,v')\text{: } d4(v,v') \in \boldsymbol{R}^{+}, v \in \boldsymbol{MS}, v' \in \boldsymbol{P}]$ - distance: warehouses - production companies,
- $\mathbf{D5} = [d5(v,v')\text{: } d5(v,v') \in \boldsymbol{R}^{+}, v \in \boldsymbol{P}, v' \in \boldsymbol{MS}]$ - distance: production companies - warehouses,
- $\mathbf{D6} = [d6(v,v')\text{: } d6(v,v') \in \boldsymbol{R}^{+}, v \in \boldsymbol{MS}, v' \in \boldsymbol{DS}]$ - distance: warehouses - suppliers,
- $\mathbf{N} = [n(veh)\text{: } n(veh) \in \boldsymbol{R}^{+}, veh \in \boldsymbol{VEH}]$ - the average pallet load units carried with one vehicle,
- $\mathbf{Q1} = [q1(v)\text{: } q1(v) \in \boldsymbol{R}^{+}, v \in \boldsymbol{DS}]$ - volume of deliveries,
- $\mathbf{Q2} = [q2(v)\text{: } q2(v) \in \boldsymbol{R}^{+}, v \in \boldsymbol{P}]$ - demand volume of enterprises,
- $\mathbf{Q3} = [q3(v,v',veh, k)\text{: } q3(v,v',veh, k) \in \boldsymbol{R}^{+}, v \in \boldsymbol{DS}, v' \in \boldsymbol{DS}, veh \in \boldsymbol{VEH}, k \in \boldsymbol{K}]$ - volume of the raw material which is transported by the vehicle between the suppliers on a given route,
- $\mathbf{Q4} = [q4(v,v',veh, k)\text{: } q4(v,v',veh, k) \in \boldsymbol{R}^{+}, v \in \boldsymbol{DS}, v' \in \boldsymbol{MS}, veh \in \boldsymbol{VEH}, k \in \boldsymbol{K}]$ - volume of the raw material which is transported by the vehicle between the supplier and the warehouse on a given route,

- **Q5** = [$q5(v,v')$: $q5(v,v') \in \boldsymbol{R}^+$, $v \in \boldsymbol{P}$, $v' \in \boldsymbol{MS}$] - volume of the raw material which belongs to the company which is stored in the given warehouse,
- **POJZ** = [$pojz(v)$: $pojz(v) \in \boldsymbol{R}^+$, $v \in \boldsymbol{MS}$] - warehouses capacity,
- ΔT - the range of time in which the collect must be realized,
- $V_{śrd}$ - the average speed,
- t_{zat} - the loading time of one pallet.

2.3 Decision Variables

The decision variables takes the forms:

- **X1** = [$x1(v,v',veh)$: $x1\,(v,v',veh) \in \{0,1\}$, $v \in \boldsymbol{P}$, $v' \in \boldsymbol{DS}$, $veh \in \boldsymbol{VEH}$] − $x1(v,v',veh)$ is equal 1 when there is a connection between the production company and the supplier which is realized by the vehicle on a given route, otherwise this variable takes the value of 0,
- **X2** = [$x2(v,v',veh,k)$: $x2\,(v,v',veh,k) \in \{0,1\}$, $v \in \boldsymbol{DS}$, $v' \in \boldsymbol{DS}$, $veh \in \boldsymbol{VEH}$, $k \in \boldsymbol{K}$] − $x2\,(v,v',veh,k)$ is equal 1 when there is a connection between the suppliers which is realized by the vehicle on a given route, otherwise this variable takes the value of 0,
- **X3** = [$x3(v,v',veh,k)$: $x3\,(v,v',veh,k) \in \{0,1\}$, $v \in \boldsymbol{DS}$, $v' \in \boldsymbol{MS}$, $veh \in \boldsymbol{VEH}$, $k \in \boldsymbol{K}$] − $x3(v,v',veh,k)$ is equal 1 when there is a connection between the supplier and the warehouse which is realized by the vehicle on a given route, otherwise this variable takes the value of 0,
- **X4** = [$x4(v,v',veh,k)$: $x4\,(v,v',veh,k) \in \{0,1\}$, $v \in \boldsymbol{MS}$, $v' \in \boldsymbol{P}$, $veh \in \boldsymbol{VEH}$, $k \in \boldsymbol{K}$] − $x4(v,v',veh,k)$ is equal 1 when there is a connection between the warehouse and the production company which is realized by the vehicle on a given route, otherwise this variable takes the value of 0,
- **X5** = [$x5(v,v',veh)$: $x5\,(v,v',veh) \in \{0,1\}$, $v \in \boldsymbol{P}$, $v' \in \boldsymbol{MS}$, $veh \in \boldsymbol{VEH}$] − $x5\,(v,v',veh)$ is equal 1 when there is a connection between the production company and the warehouse which is realized by the vehicle on a given route, otherwise this variable takes the value of 0,
- **X6** = [$x6(v,v',veh,k)$: $x6\,(v,v',veh,k) \in \{0,1\}$, $v \in \boldsymbol{MS}$ $v' \in \boldsymbol{DS}$, $veh \in \boldsymbol{VEH}$, $k \in \boldsymbol{K}$] − $x6\,(v,v',veh,k)$ is equal 1 when there is a connection between the warehouse and the supplier which is realized by the vehicle on a given route, otherwise this variable takes the value of 0.

2.4 Constraints

The constraints take the following forms:

- the capacity of the vehicle cannot be exceeded:

$$\begin{gathered}\forall \mathrm{veh} \in VEH\ \forall \mathrm{k} \in K\\ \sum_{\mathrm{v}\in \mathrm{DS}}\ \sum_{\mathrm{v}'\in \mathrm{DS}:\mathrm{v}\neq \mathrm{v}'} \mathrm{x2(v,v',veh,k)}\cdot \mathrm{q3(v,v',veh,k)}+\\ +\sum_{\mathrm{v}\in \mathrm{DS}}\ \sum_{\mathrm{v}'\in \mathrm{MS}} \mathrm{x3(v,v',veh,k)}\cdot \mathrm{q4(v,v',veh,k)}\leq \mathrm{n(veh)}\end{gathered} \tag{1}$$

- the capacity of the warehouses cannot be exceeded:
- for one vehicle:

$$\forall v \in \boldsymbol{MS}, \forall veh \in \boldsymbol{VEH} \sum_{v' \in \boldsymbol{DS}} \sum_{k \in \boldsymbol{K}} x3(v, v', veh, k) \cdot q4(v, v', veh, k) \leq pojz(v) \tag{2}$$

– for vehicles:

$$\forall v \in \boldsymbol{MS} \sum_{v' \in \boldsymbol{DS}} \sum_{veh \in \boldsymbol{VEH}} \sum_{k \in \boldsymbol{K}} x3(v, v', veh, k) \cdot q4(v, v', veh, k) = pojz(v) \tag{3}$$

- the demand of companies must be met:

$$\forall v \in \boldsymbol{P} \sum_{v' \in \boldsymbol{MS}} \sum_{veh \in \boldsymbol{VEH}} \sum_{k \in \boldsymbol{K}} x5(v, v', veh, k) \cdot q5(v, v') = q2(v) \tag{4}$$

- the production capacity of suppliers cannot be exceeded, the vehicle can go to the other supplier or to the warehouses many times:

$$\forall v \in \boldsymbol{DS}$$

$$\sum_{v' \in \boldsymbol{DS}: v \neq v'} \sum_{veh \in \boldsymbol{VEH}} \sum_{k \in \boldsymbol{K}} x2(v, v', veh, k) \cdot q3(v, v', veh, k) + \sum_{v' \in \boldsymbol{MS}} \sum_{veh \in \boldsymbol{VEH}} \sum_{k \in \boldsymbol{K}} x3(v, v', veh, k) \cdot q4(v, v', veh, k) \leq q1(v) \tag{5}$$

- the delivery time to warehouses must be met:

When the beginning of the route is in the companies, the route consists of the segment between the production company and the supplier (only one because the vehicle passes through this segment only once when it starts the route), the segments between the suppliers (the vehicle can pass through those segments several times) and the segments between suppliers and the warehouse (the vehicle can pass through those segments several times):

$$\forall v \in \boldsymbol{MS},\ \forall veh \in \boldsymbol{VEH}, \forall v' \in \boldsymbol{P}, \forall v'' \in \boldsymbol{DS}$$

$$\frac{x1(v',v'',veh,k)\cdot d1(v',v'')}{v_{śrd}}+\sum_{v'''\in \boldsymbol{DS}}\sum_{v''''\in \boldsymbol{DS}}\sum_{k\in \boldsymbol{K}}x2(v''',v'''',veh,k)\cdot\left[q3(v''',v'''',veh,k)\cdot t_{zal}+\frac{d1(v''',v'''')}{v_{śrd}}\right]$$
$$+\sum_{v'''\in \boldsymbol{DS}}\sum_{k\in \boldsymbol{K}}x3(v''',v,veh,k)\cdot\left[q4(v''',v,veh,k)\cdot t_{zal}+\frac{d3(v''',v)}{v_{śrd}}\right]\leq\Delta T \tag{6}$$

When the beginning of the route is in the warehouses, the route consists of the segments between warehouses and suppliers (the vehicle can pass through those segments several times), the segments between the suppliers (the vehicle can several times pass through those segments) and the segments between suppliers and the warehouse (the vehicle can pass through those segments several times):

$$\forall v\in \boldsymbol{MS},\ \forall veh\in \boldsymbol{VEH}$$

$$\sum_{v'\in \boldsymbol{DS}}\sum_{k\in \boldsymbol{K}}\frac{x6(v,v',veh,k)\cdot d6(v,v')}{v_{śrd}}+\sum_{v'\in \boldsymbol{DS}}\sum_{v''\in \boldsymbol{DS}}\sum_{k\in \boldsymbol{K}}x2(v',v'',veh,k)\cdot\left[q3(v',v'',veh,k)\cdot t_{zal}+\frac{d1(v',v'')}{v_{śrd}}\right]$$
$$+\sum_{v'\in \boldsymbol{DS}}\sum_{k\in \boldsymbol{K}}x3(v',v,veh,k)\cdot\left[q4(v',v,veh,k)\cdot t_{zal}+\frac{d3(v',v)}{v_{śrd}}\right]\leq\Delta T \tag{7}$$

2.5 The Criterion Function

The criterion function looks for the minimal routes in which the raw material is carried out to the production companies. The minimal routes depends on the location of warehouses in the logistic network. The criterion function takes the following forms:

$$F(X1,X2,X3,X4,X5,X6)=$$
$$\sum_{v\in P}\sum_{v'\in DS}\sum_{veh\in VEH}\sum_{k\in K}x1(v,v',veh,k)\cdot d1(v,v')+\sum_{v\in DS}\sum_{v'\in DS}\sum_{veh\in VEH}\sum_{k\in K}x2(v,v',veh,k)\cdot d2(v,v')$$
$$+\sum_{v\in DS}\sum_{v'\in MS}\sum_{veh\in VEH}\sum_{k\in K}x3(v,v',veh,k)\cdot d3(v,v')+\sum_{v\in MS}\sum_{v'\in P}\sum_{veh\in VEH}x4(v,v',veh)\cdot d4(v,v')$$
$$+\sum_{v\in MS}\sum_{v'\in DS}\sum_{veh\in VEH}\sum_{k\in K}x6(v,v',veh,k)\cdot d6(v,v')+2\cdot\sum_{veh\in VEH}\sum_{v\in P}\sum_{v'\in MS}\left\lceil\frac{q5(v,v')}{n(veh)}\right\rceil\cdot x5(v,v',veh)\cdot d5(v,v')$$
$$\longrightarrow\min \tag{8}$$

3 The Ant Algorithm in the Warehouse Location - Routing Problem

In order to solve the warehouses location problem taking into account the vehicle routing problem, the ant algorithm was developed. Theory of this algorithms introduces the concept of artificial ants [17]. The main task of ants is to determine the minimum delivery routes of raw materials to the companies. The value of the ant must be interpreted as the length of the route. Each ant builds its own route till the moment when the demand of companies will be met or the time constraints (6–7) will not be met. Then the ant returns to the company. Each ant has its own capacity which is equal to the capacity of the vehicle. At first the data input needs to be determined: the set of ants was defined as ***MR***, the number of iteration of the algorithm ***I***. The starting and ending point of each route is in the production company. Step 1 of the ant algorithm is to select the production company in a random way, the first ant and the first iteration of the algorithm is set. Step 2 is to determine the initial value of the capacity of warehouses, the demand of companies. The ant goes to the subsequent points (e.g. suppliers, warehouses) with a certain probability, step 3:

$$PR_{yz}^{mr}(t) = \begin{cases} \frac{[\tau_{yz}(t)]^{\alpha} \cdot [\eta_{yz}(t)]^{\beta}}{\sum_{l \in \Omega^{mr}} [\tau_{yl}(t)]^{\alpha} \cdot [\eta_{yl}(t)]^{\beta}}, & z \in \Omega^{mr} \\ 0, & z \notin \Omega^{mr} \end{cases} \tag{9}$$

where:

- $\tau_{yz}(t)$ - the intensity of pheromone trail between the y -the a z -the point in t -the iteration,
- $\eta_{yz}(t)$ - the heuristic information e.g. $\eta_{yz}(t) = 1/w(y, z)$, where $w(y,z)$ - the distance between y -the point of route and z -the point of the route,
- α, β - parameters determining the effect of pheromones and the heuristic information on the behavior of ants.
- Ω^{mr} - the set of l - points to the route which has not been yet visited by mr- this ant.

Random choice of the route between the point y and points l begins from calculating the probability of the transition to the points which have not yet been visited by the ant according to the pattern (9). The next step is to calculate the distribution for each transition path and draw the number r from the range [0,1]. The route tr about the value of the distribution q_{tr} which fulfills the condition $q_{tr-1} < r \leq q_{tr}$ is selected, where tr is the number of the route between y - the point and l - the points. At the beginning the ant considers only the points which are defined as suppliers. The moment when the ant is fully loaded the only points which are defined as warehouses are considered in the further route.

In step 4 the ant collects the raw material from suppliers taking into account the constraint: the capacity of the vehicle cannot be exceeded (1) and the production capacity of suppliers cannot be exceeded (5), the time constraint (6). The value of the collected raw material is adequate to the demand of the production company. At this point the ant has two options: when the ant is fully loaded it goes to the warehouse or it goes to the suppliers when its capacity is not fully used. The same situation takes place in the case of the time constraint. When the time constraint is fulfilled the ant goes to the suppliers, otherwise the ant goes to the warehouse.

In step 5 the probability of the transition to the warehouse is calculated. In step 6 the ant unloads the raw material in the warehouse. In case when the capacity of the warehouse is not fulfilled another warehouse is selected, step 7. Otherwise, the constraint of the demand of the company is checked.

In case when the demand of the company is met the ant returns to company. The route of the ant is finished, step 8. In step 9 the new routes are determined for the same company. In case when the demand of the company is not met, the time constraints are checked. If the time constraints are fulfilled the ant goes to step 2, if this constraint is not fulfilled the additional vehicle on the route is introduced and the ant goes to step 2, step 10. At the beginning it is assumed that each route of the ant is realized by one vehicle. The route is again designated. In step 11 all routes are constructed. Those routes are correct for all the constraints. In step 12 the pheromone update is actualized. Three types of the pheromone update can be distinguished: ant - density, ant - quantity, ant - cycle. In order to update the pheromone the ant - cycle was used as the most efficient version of the ant algorithm [17].

At the beginning it is assumed that the trail on the links between the points is equally strong. In subsequent iterations, the pheromone trail is calculated according to the formula:

$$\tau_{yz}(t+1) = (1-\rho)\tau_{yz}(t) + \sum_{mr=1}^{MR} \Delta\tau_{yz}^{mr}(t) \tag{10}$$

where:

mr - another ant $mr \in \boldsymbol{MR}$ mr $\in$ **MR**,
ρ - a factor pheromone ($0 < \rho \leq 1$),
$\tau_{yz}(t + 1)$ - the strengthening of the pheromone, for the first iteration this strengthening takes the value τ_0 for each connections.

$$\Delta\tau_{yz}^{mr}(t) = \begin{cases} \frac{1}{L^{mr}(t)} & \textit{when the route } (y,z) \textit{ was used by mr} - \textit{this ant} \\ 0 & \textit{otherwise} \end{cases} \tag{11}$$

where:

$L^{mr}(t)$ - the length of the route in t iteration realized by mr,
mr - this ant, if the segment of routes (y,z) was realized by mr - this ant then $\Delta\tau_{yz}^{mr}(t)$ equals $1/L^{mr}(t)$, otherwise 0.

The ant algorithm was presented in Fig. 1.

The ant algorithm is repeated a given number of the iteration ***I***. The final effect of work of ants are the routes used in order to transport the raw material to the production company. Among those routes one can select the route of the minimum value which is the solution, step 13. For other companies this procedure goes the same way.

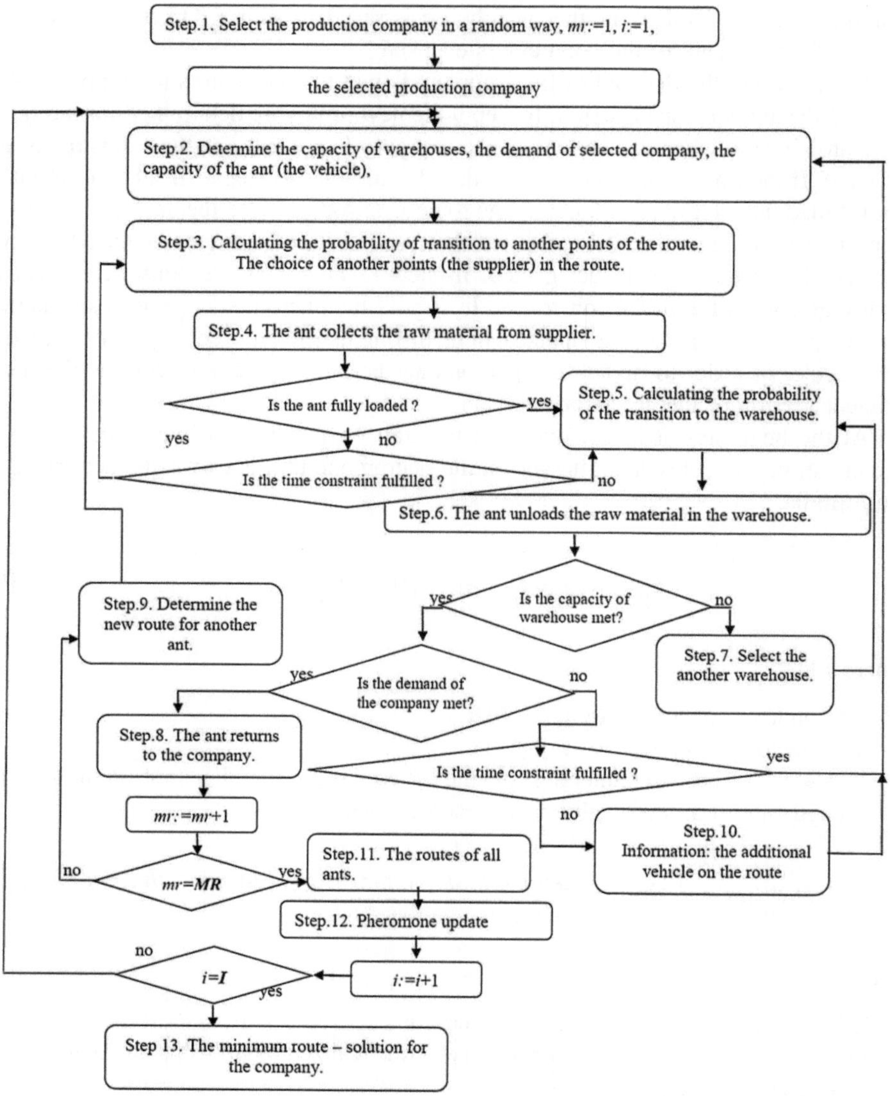

Fig. 1. The ant algorithm for the warehouse location - routing problem

4 Implementation of Ant Algorithm

The algorithm was implemented in C language programming by the use of the real input data e.g. Table 1 - the production capacity of suppliers, the capacity of warehouses - Table 2, the demand of the production company - 150 pallet units, the average pallet load units carried with one vehicle - 30 units.

Table 1. The production capacity of suppliers.

Suppliers	Location	Unit
1	Warszawa	10
2	Siedlce	10
3	Wrocław	20
4	Poznań	30
5	Piła	20
6	Gdynia	30
7	Kielce	40
8	Opole	30
9	Lublin	40
10	Gdańsk	20
11	Ryki	40
12	Łuków	15
13	Parczew	15
14	Płońsk	20
15	Katowice	10
16	Ełk	10

Table 2. The capacity of warehouses.

Warehouses	Location	Capacity/unit
1	Kraków	100
2	Mińsk Mazowiecki	200
3	Międzyrzec Podlaski	300
4	Radzyń Podlaski	50
5	Ciechanów	230
6	Ostrołęka	140
7	Lubartów	200
8	Mińsk Mazowiecki	150

The first step of implementation of the ant algorithm was to find the set of the best parameters which characterizes this algorithm. The following combinations of parameters and results of tests were presented in Table 3.

Based on the results from Table 3, the optimal parameters were designated (test 3: $\alpha = 1$, $\beta = 5$, $\rho = 0{,}2$). The graphical representation of the results of tests of parameters was shown on Fig. 2.

In order to verify the correctness of the ant algorithm (AA), its results were compared with random values (AL). The ant algorithm in each case generated a better solution than the random algorithm. The results are shown in Tables 3 and 4.

Table 3. Combination of parameters for the ant algorithm.

Test/routes [km]	α	β	ρ
1/2530	1	0.2	0.2
2/1470	1	0.5	0.8
3/850	1	5	0.2
4/2690	1	1	0.8
5/1280	3	0.5	0.2
6/1700	3	0.5	0.8
7/3000	3	5	0.2
8/2896	3	5	0.8
9/2908	5	5	0.2
10/3410	5	5	0.8
11/2700	10	5	0.2
12/1641	10	5	0.8
13/1670	20	0.5	0.2
14/1700	20	5	0.8

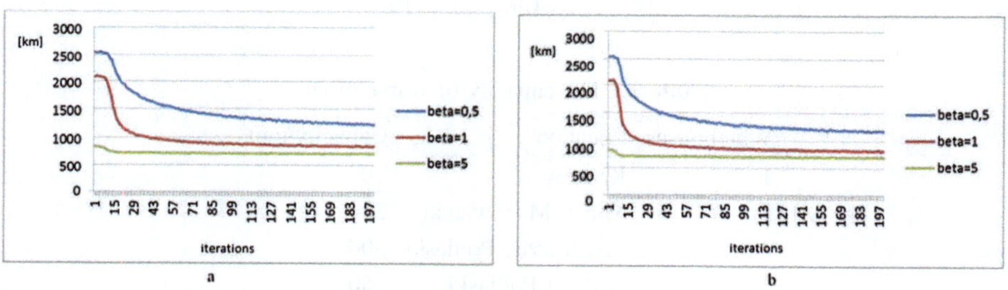

Fig. 2. The example tests of the ant algorithm

Table 4. Combinations of parameters for the ant algorithm.

Ant algorithm [km]	Random algorithm [km]	Ant algorithm [km]	Random algorithm [km]
780	1700	920	1100
820	1820	1100	1700
730	2100	820	2100
1200	2500	1430	2300
1020	2300	970	1670
980	2500	1300	1903
1100	2347	1270	1700
1304	1908	1002	1980

5 Conclusion

The optimal results generated by the algorithm depend on many factors e.g.: parameters of the algorithm, the number of iterations or ants in population. The selection of these parameters takes place in a random way. In the implementation of the ant algorithm the different combinations of these parameters were examined i.e.: 100 iterations. It should be emphasized that the solution generated by the algorithm is sub - optimal. In spite of this, the speed of the algorithm is its advantage. Further research in the context of the use of the ant algorithm in the location of warehouses in the logistic network will be conducted using anti-colonial systems described in the literature.

It should be underlined that results of the ant algorithm depend on the type of the data input which are took into account in the mathematical model. The ant algorithm was implemented by the use of the fixed data input e.g. the average speed or the loading time of one pallet. The presented model does not take into account the random values of the input data. Further research in the context of warehouse location-routing problem will be conducted taking into account random character of transport process e.g. random input data can be generated by the use of the artificial neural network.

The developed algorithm can be used in production companies to develop the driver's working schedules, the minimal number of vehicles which realize given transportation tasks. This algorithm cannot be used in the production companies in which time windows of delivery are determined. The main advantage of the ant algorithm is that the results are generated in a quickly way, what is very important for the production companies. The processes occurring in the production companies are the dynamic processes. For this reason the ant algorithm must be started a few times depending on recipients demands, warehouses capacities. In this case the calculation speed plays the huge role what underlines utility of this algorithm in the production companies.

Acknowledgements. The scientific work carried out in the frame of PBS 3 project "System for modelling and 3D visualization of storage facilities" (SIMMAG 3D) financed by the National Centre for Research and Development.

References

1. Sharma, R.R.K., Berry, V.: Developing new formulations and relaxations of single stage capacitated warehouse location problem: empirical investigation for assessing relative strengths and computational effort. Eur. J. Oper. Res. **177**(2), 803–812 (2007)
2. Khumawala, B.M.: An efficient heuristic procedure for the uncapacitated warehouse location problem. Nav. Res. Logist. Q. **20**(1), 109–121 (1973)
3. Akinc, U., Khumawala, B.M.: An efficient branch and bound algorithm for the capacitated warehouse location problem. Manag. Sci. **23**(6), 585–594 (1977)
4. Jacyna-Gołda, I.: Chosen aspects of logistics network design method for production service companies. Int. J. Logist. Syst. Manag. **15**(2/3), 219–238 (2013)
5. Wasiak, M., Jacyna-Gołda, I., Izdebski, M.: Multi-criteria warehouses location problem in the logistics network. In: Proceedings of International Conference on Industrial Logistics, pp. 352–363 (2016)

6. Dey, B., Bairagi, B., Sarkar, D., Sanyal, S.K.: Warehouse location selection by fuzzy multi-criteria decision making methodologies based on subjective and objective criteria. Int. J. Manag. Sci. Eng. Manag. **11**, 262–276 (2015)
7. Jacyna, M.: Multicriteria evaluation of traffic flow distribution in a multimodal transport corridor, taking into account logistics base service. Arch. Transp. **10**(1–2), 43–66 (1999)
8. Jacyna, M., Wasiak, M.: Multicriteria decision support in designing transport systems. In: Mikulski, J. (ed.) Tools of Transport Telematics. CCIS, vol. 531, pp. 11–23. Springer, Heidelberg (2015)
9. Zieja, M., Smoliński, H., Gołda, P.: Information systems as a tool for supporting the management of aircraft flight safety. Arch. Transp. **36**(4), 67–76 (2015)
10. Jacyna-Gołda, I., Izdebski, M., Szczepański, E.: Assessment of the method effectiveness for choosing the location of warehouses in the supply network. In: Mikulski, J. (ed.) Challenge of Transport Telematics. CCIS, vol. 640, pp. 84–97. Springer, Heidelberg (2016)
11. Demirel, T., Demirel, N., Kahraman, C.: Multi-criteria warehouse location selection using choquet integral. Expert Syst. Appl. **37**, 3943–3952 (2010)
12. Hansen, P.H., Hegedahl, B., Hjortkjaer, S., Obel, B.: A heuristic solution to the warehouse location-routing problem. Eur. J. Oper. Res. **76**(1), 111–127 (1994)
13. Perl, J., Daskin, M.S.: A warehouse location-routing problem. Transp. Res. Part B **19**(5), 381–396 (1985)
14. Izdebski, M.: The use of heuristic algorithms to optimize the transport issues on the example of municipal services companies. Arch. Transp. **29**(1), 27–36 (2014)
15. Dorigo, M., Gambardela, L.M.: Ant colonies for the travelling salesman problem. BioSystems **43**, 73–81 (1997)
16. Dorigo, M., Gambardela, L.M.: Ant colony system: a cooperative learning approach to the traveling salesman problem. IEEE Trans. Evol. Comput. **1**(1), 53–66 (1997)
17. Dorigo, M., Stutzle, T.: Ant Colony Optimization. Bradford Books, USA (2004)

Safety Analysis in Advanced Transport Networks and Transport Systems

Road Infrastructure Condition Assessment as Element of Road Traffic Safety – Concept of the RCT Solution in the S-mileSys Platform

Marcin Staniek(✉)

Faculty of Transport, Silesian University of Technology, Katowice, Poland
marcin.staniek@polsl.pl

Abstract. The paper provides a discussion on the overall body of problems related to developing a mobile tool referred to as Road Condition Tool, used for the road infrastructure condition assessment based on the recorded signals of linear accelerations set out in a three-dimensional system. Such a data set is created while typical transport services are performed by a group of carriers which have deployed the solution in question. The road condition tool discussed in this article constitutes a part of the S-mileSys platform developed under the "Sustainable Logistics and Supply Chains" competition within the framework of the ERANET Transport III programme. The solution proposed in the paper is a mobile device with a module for positioning and one for recording linear accelerations, featuring software developed under the S-mile project for purposes of the road network assessment and description.

Keywords: Road pavement assessment · Road profile · Roughness · Vibrations · Road traffic safety

1 Introduction

Road infrastructure comprises stationary road structures used for transportation of passengers and cargo by different means of transport as well as by commuting on foot. It forms a network which consists of junctions and junction links, where a junction is not only understood as a multi-level road junction, but also as an at-grade road intersection assuming all the possible forms, i.e. intersections with and without traffic lights or roundabouts. Junction links are roads and streets connecting individual road junctions [1–3].

In other words, road infrastructure is a key element of the vehicle-man-road system whose role in the safe, economic and social life is of major importance. It provides grounds for rendering transport services, but at the same time a modern system of transport must be sustainable and satisfy all preferences and requirements of its participants [4–6]. It is additionally required to meet the criteria of minimisation of negative impact on natural environment and human health, which is particularly emphasised in numerous reference publications, including strategic documents of the European Union [7–9].

E. Macioszek and G. Sierpiński (eds.), *Recent Advances in Traffic Engineering for Transport Networks and Systems*, Lecture Notes in Networks and Systems 21,
DOI 10.1007/978-3-319-64084-6_6

2 Solutions for Road Infrastructure Condition Assessment

Since road infrastructure is in fact a technical structure, it requires periodical technical condition inspections and detection of defects aimed at elimination of potential threats to road traffic safety through identification and execution of suitable repair procedures or works [10]. At the same time, in accordance with Directive 2008/96/EC of the European Parliament and of the Council of 19 November 2008 on road infrastructure safety management, one is required to apply procedures of road traffic safety (RTS) control including the impact of technical condition on the level of RTS. According to guidelines of the Pavement Management System (PMS), technical and operating parameters subject to assessment under the road infrastructure diagnostic procedure include the following:

- load bearing capacity of structures,
- longitudinal evenness,
- transverse evenness,
- pavement condition (identification of surface defects),
- anti-skid properties.

One can use a number of methods to determine the load bearing capacity of a road structure, including the pressure plate method and the deflection method. The basic assumption is that road samples collected from a test section are subject to strength tests, based on which one determines the modulus of elasticity and the stresses emerging in road pavement. For purposes of non-invasive measurements of deflections, one can use a falling weight deflectometer (FWD) [10] or the latest laser-based solution, namely a traffic speed deflectometer (TSD). With regard to TSD, tests are conducted in a contactless manner, where the measurement of the road pavement displacements (under the impact of the test trailer wheel) is conducted using a laser range finder. The device is installed in a tractor unit's semi-trailer [11].

Transverse and longitudinal evenness are further technical and operating parameters which describe the condition of road infrastructure. One of the instruments used in contemporary diagnostics are laser profilographs characterised by high precision and simplicity of measurements. They enable continuous measurement of the road profile conducted under conditions of real traffic in the road network, causing no additional hindrance and, even more importantly, posing no threat to road traffic safety.

Identification and recording of surface defects constitute the next step in the infrastructure condition assessment, which also includes surveying of bumps, deformations, chipping defects and patches as well as various fractures, including longitudinal, transverse, alligator-type and technological cracks. For the sake of identification of the foregoing types of defects, one can apply visual solutions based on image processing and recognition technologies. Such measures as morphological operations, segmentation, object detection or binarisation provide major support in the assessment process, however, in order to be used, they require monitoring and verification by a video research material inspector on account of the considerable imperfections of the visual method which result, for instance, from non-homogeneous and dynamic illumination, discontinuity of road pavement mapping, particularly when stereo vision methods are used, or when objects obscure the view of one another.

In order to estimate the anti-skid properties of a road, one must determine the friction coefficient which corresponds to road adhesion [12]. One of the methods that can serve that purpose is using a measuring set equipped with a dynamometric wheel, where the analysis of the test results obtained by having initiated a braking action determines the grade of road pavement slipperiness. An alternative solution is testing the road pavement texture by means of pre-calibrated sand spread over the surface tested, where the assumed sand quantity and its spread determine the roughness of the pavement examined. As provided in the literature of the subject [13, 14], there are also optional visual methods and laser technologies that can be used to identify pavement roughness.

Test vehicles, also referred to a mobile laboratories, are solutions which enable assessment of multiple technical and operating parameters of a road in a single measuring run. A set of parameters thus obtained, describing the technical condition of road infrastructure, comprises assessments and records of longitudinal and transverse evenness, but also makes it possible to identify surface defects, such as:

- bumps,
- deformations,
- chipping defects,
- patches,
- cracks.

Moreover, in the course of the measuring run, the vehicle's surroundings are recorded in a sequence of photographs showing individual elements of the roadway infrastructure. Sample mobile laboratories are vehicles like ARGUS, IBDiM or GREENWOOD [15, 16].

Another solution applied to assess the road infrastructure condition is to measure the linear accelerations recorded while a vehicle is running on a road. The basic method consists in using an accelerometer to record vibrations of a wheel loaded with a mass of pre-set weight as it is moving on the road being examined, however, an analysis of the results thus obtained requires that the noise caused by damping of the mass system be filtered out of the recorded signal. Individual solutions based on the foregoing approach have been described in papers [17–26]. The solutions addressed in these publications have set the orientation for further development of a concept of a tool for the road infrastructure condition assessment based on an analysis of linear acceleration signals in a three-dimensional description.

3 Describing Quality of Transport Roads in the S-mileSys System

As an outcome of international collaboration between institutions representing three countries: Spain, Turkey and Poland, a research project was initiated with the aim of developing a "Smart platform to integrate different freight transport means, manage and foster first and last mile in supply chains (S-mile)", co-financed under the "Sustainable Logistics and Supply Chains" competition within the framework of the ERANET

Transport III programme. The S-mileSys platform, currently being developed, is an integrated IT system intended to be used by such parties as clients searching for local freight operators, freight companies and local authorities as well as road infrastructure administration bodies. The S-mileSys platform architecture is based on six fundamental tools/modules which perform the following functions:

- integration of freight transport companies,
- searching for first/last mile carriers,
- vehicle fleet monitoring,
- single-run distribution of goods (covering distribution and collection),
- route planning by taking the road infrastructure condition, road traffic conditions and new/updated orders into consideration,
- route optimisation based on primary criteria of distance, time and cost,
- route optimisation in line with innovative criteria: social and environmental impact,
- creating reports, analyses and statistics to calculate total costs of freight transport in the system,
- assessments of the environmental impact of various motivational/fiscal initiatives, timetable changes, parking areas etc. exerting influence on how commercially, socially and environmentally sound freight transport is managed,
- monitoring of the transport system condition and acquisition of information on the freight transport taking place in the chosen section of the road network.

The S-mileSys system architecture diagram highlighting the aforementioned tools/modules has been provided in Fig. 1.

Fig. 1. Tools of S-mileSys

In accordance with the assumptions underlying the S-mileSys platform being developed, the road infrastructure condition, i.e. the quality of transport roads, is a crucial parameter which should be taken into account while planning and optimising routes for both delivery and collection of goods. However, this requires data on the road infrastructure condition (technical and operating parameters) to be obtained and subsequently set out in a graph of connections which represents the road network. The graph of connections is a key component of S-mile Transport Planner Tool used to perform the procedure of freight transport route planning.

The infrastructure describing data may be acquired for purposes of the S-mileSys system in two ways, i.e. one can make use of the PMS data or collect data while running road pavement condition tests. However, it is possible that the PMS data do not cover the entire road network, as they typically pertain to roads of higher categories and remain up-to-date for only a short period of time directly following measurements. On the other hand, measuring technical and operating parameters of the road network for the sake of their assessment involves high costs, particularly when the solutions referred to in the previous section of the article are used. Bearing the foregoing in mind, specific solutions have been proposed in the S-mileSys system for the road pavement condition assessment on the basis of an analysis of linear accelerations recorded by fleet vehicles using the system in question in day-to-day operations.

4 Requirements Towards Road Condition Tool

Road Condition Tool (RCT) is a mobile device featuring a dedicated application implemented in the Android operating system, intended for recording of vehicle motion parameters and the shocks resulting from the road pavement condition. From the infrastructure assessment perspective, the RCT application installed on the mobile device has been designed to record data of:

- linear accelerations in a three-dimensional system, using the MEMS module,
- position in the WGS'84 system, using the GPS module.

The above data are stored in the mobile device's memory, only to be sent to the RCT data server using GSM data transmission or a wireless network. The server receives and archives the data extracted from the mobile devices, and then performs the procedures of input data processing, including filtration and aggregation.

As individual transfers are made, the RCT application generates an output file which contains the following information:

1. Mandatory header fields:
    - id company; id route; vehicle type; weight of freight; weight of vehicle - data concerning the given transfer being made, required for data processing in the S-mileSys system,
    - step - read-out interval for successive data in the sequence of linear accelerations recorded,
    - tripStart - date and time of measurement start which corresponds to the freight transport commencement.

2. Series of output data comprising the vehicle motion parameters and shocks recorded during transfers. The detailed description includes: latitude [WGS'84]; longitude [WGS'84]; X linear acceleration [m/s^2]; Y linear acceleration [m/s^2]; Z linear acceleration [m/s^2]; speed [km/h].

A sample of recorded linear accelerations along with a reference to the road network has been shown in Fig. 2. Highlighted in colours are the junction linking sections which break the recorded linear acceleration signal down into fragments of the graph illustrating the passes of the S-mile Transport Planner Tool, which proves critical when planning routes by taking the road infrastructure condition into consideration.

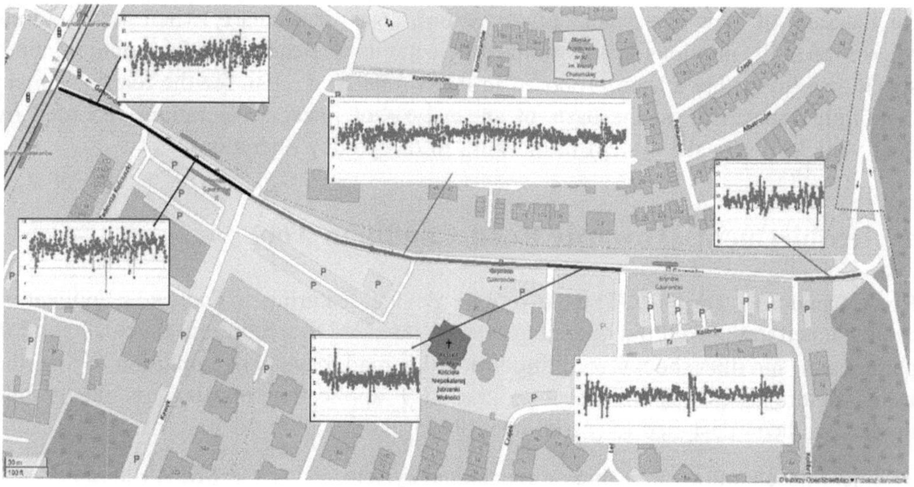

Fig. 2. Linear accelerations recorded along the chosen route

5 Conclusions

A review of reference publications addressing the solutions used for purposes of the road pavement condition assessment, and primarily those that rely on an analysis of vibrations observed during vehicle motion, has implied that the approach assumed for the Road Condition Tool concept is correct. The linear acceleration signal recorded in a hands-off mode in the course of freight transport enables identification of point and section based defects of road pavement. An additional analysis of the data thus obtained makes it possible to determine the scope and magnitude of the road pavement degradation.

Having analysed the potential of the recorded data set, one may conclude that it is possible to extend the road infrastructure condition assessment with a determinant of quality of the transport services rendered by drivers employed at the given freight company. What it also enables is identification of transport and waiting times, the driver's driving style vs. road traffic safety as well as the level of security of the cargo being transported.

Acknowledgements. The present research has been financed from the means of the National Centre for Research and Development as a part of the international project within the scope of ERA-NET Transport III Sustainable Logistics and Supply Chains Programme "Smart platform to integrate different freight transport means, manage and foster first and last mile in supply chains (S-mile)".

The National Centre for Research and Development

S mile

References

1. Macioszek, E.: The influence of motorcycling and cycling on small one-lane roundabouts capacity. In: Mikulski, J. (ed.) Modern Transport Telematics. CCIS, vol. 239, pp. 291–298. Springer, Heidelberg (2011)
2. Szczuraszek, T., Macioszek, E.: Proportion of vehicles moving freely depending on traffic volume and proportion of trucks and buses. Balt. J. Road Bridge Eng. **8**(2), 133–141 (2013)
3. Macioszek, E., Sierpiński, G., Czapkowski, L.: Methods of modeling the bicycle traffic flows on the roundabouts. In: Mikulski, J. (ed.) Transport Systems Telematics. CCIS, vol. 104, pp. 115–124. Springer, Heidelberg (2010)
4. Sierpiński, G.: Model of incentives for changes of the modal split of traffic towards electric personal cars. In: Mikulski, J. (ed.) Telematics - Support for Transport. CCIS, vol. 471, pp. 450–460. Springer, Heidelberg (2014)
5. Sierpiński, G., Staniek, M., Celiński, I.: Research and shaping transport systems with multimodal travels – methodological remarks under the green travelling project. In: Proceedings of the 7th International Conference of Education. Research and Innovation ICERI Proceedings, pp. 3101–3107. IATED, Seville (2014)
6. Sierpiński, G.: Revision of the modal split of traffic model. In: Mikulski, J. (ed.) Activities of Transport Telematics. CCIS, vol. 395, pp. 338–345. Springer, Heidelberg (2013)
7. Commission, European: Clean Power for Transport: A European Alternative Fuels Strategy. European Commission, Brussels (2013)
8. European Commission: Action Plan on Urban Mobility. European Commission, Brussels (2009)
9. Commission, European: White Paper: Roadmap to a Single European Transport Area - Towards a Competitive and Resource Efficient Transport System. European Commission, Brussels (2011)
10. Sybylski, D., Mechowski, T., Harasim, P.: Evaluation of effectiveness of use of FWD for assessment of pavement interlayer bond. Roads Bridges **6**, 41–83 (2007)
11. Chambon, S., Moliard, J.M.: Automatic road pavement assessment with image processing: review and comparison. Int. J. Geophys. **2011**, 1–20 (2011)
12. Flintsch, G.W.: The Little Book of Tire Pavement Friction. Pavement Surface Properties Consortium, USA (2012)
13. El Khoury, J., Akle, B., Katicha, S., Ghaddar, A., Daou, M.: A microscale evaluation of pavement roughness effects for asset management. Int. J Pavement Eng. **15**(4) (2014)
14. Sayers, M.W., Gillespie, T.D., Peterson, W.D.O.: Guidelines for Conducting and Calibrating Road Roughness Measurements. The International Bank for Reconstruction and Development, USA (1986)

15. Yu, S., Sukumar, S.R., Koschan, A.F., Page, D.L., Abidi, M.A.: 3D Reconstruction of road surfaces using an integrated multi-sensory approach. Opt. Lasers Eng. **45**(7), 808–818 (2007)
16. He, Y., Wang, J., Qiu, H., Zhang, W., Xie, J.: A research of pavement potholes detection based on three-dimensional projection transformation. In: Proceedings of the 4th International Congress on Image and Signal Processing, Shanghai, pp. 1805–1808 (2011)
17. Burdzik, R., Celiński, I., Czech, P.: Optimization of transportation of unexploded ordnance, explosives and hazardous substances - vibration issues. Vibroeng. Proc. **10**, 382–386 (2016)
18. Młyńczak, J., Burdzik, R., Celiński, I.: Research on dynamics of shunting locomotive during movement on marshalling yard by using prototype of remote control unit. In: Awrejcewicz, J. (ed.) Dynamical Systems: Theoretical and Experimental Analysis. SPMS, vol. 182, pp. 279–292. Springer, Berlin (2016)
19. Celiński, I.: Transport network parameterisation using the GTAlg tool. In: Macioszek, E., Sierpiński, G. (eds.) Contemporary Challenges of Transport Systems and Traffic Engineering. LNNS, vol. 2, pp. 111–123. Springer, Switzerland (2017)
20. Sijing, G., Yilun, L., Lin, X., Xuexun, G., Lei, Z.: Performance evaluation and parameter sensitivity of energy-harvesting shock absorbers on different vehicles. Veh. Syst. Dyn. **54**, 9–18 (2016)
21. Mukherjee, A., Subhra, M.: Characterisation of road bumps using smartphones. Eur. Transp. Res. Rev. **8**, 1–12 (2016)
22. Ghose, A., Biswas, P., Bhaumik, C., Sharma, M., Pal, A., Jha, A.: Road condition monitoring and alert application: using in-vehicle smartphone as internet-connected sensor. In: Pervasive Computing and Communications Workshops, pp. 489–91. IEEE Press, Lugano (2012)
23. Agostinacchio, M., Ciampa, D., Olita, S.: The vibrations induced by surface irregularities in road pavements - a Matlab approach. Eur. Transp. Res. Rev. **6**, 267–275 (2014)
24. Melcer, J.: Vehicle-road interaction, analysis in a frequency domain. Slovak J. Civ. Eng. **3**, 48–52 (2006)
25. Andren, P.: Power spectral density approximations of longitudinal road profiles. Int. J. Veh. Des. **40**, 2–14 (2006)
26. Yousefzadeh, M., Azadi, S., Soltani, A.: Road profile estimation using neural network algorithm. J. Mech. Sci. Technol. **24**, 743–754 (2010)

Safe Routes as One of the Ways to Reduce the Number of Road Accidents Victims

Irina Makarova[1(✉)], Anton Pashkevich[2], and Ksenia Shubenkova[1]

[1] Kazan Federal University, Naberezhnye Chelny, Russia
kamivm@mail.ru, ksenia.shubenkova@gmail.com
[2] Cracow University of Technology, Krakow, Poland
anton.pashkevich@gmail.com

Abstract. The article deals with actual problems pedestrians and cyclists safety in the cities. It is shown that, despite the positive experience of interventions implementation, road traffic deaths and injuries are still high. One of the possible ways to ensure road safety for the most vulnerable road users is to develop the method to support the safest route selection. The proposed method will increase the quality of management and effectiveness of the adopted solutions to improve traffic safety.

Keywords: Road safety management · Route selection · Pedestrian safety · Bicycle transport

1 Introduction

The scale and significance of problems associated with positive changes in image of the world transport in the 21st century, are assessed as strategic challenges to national and even continental scale. First of all, it concerns the urbanized territories where the negative impacts associated with the development and functioning of the transport complex reduce the quality of people's life. This is true also for the reason that by 2030 the majority of the world's population will be concentrated in cities. Assuming this trend continues, by 2050 more than 80% of the world's population will live in an urban environment. Therefore, using the benefits of the transport system in the area of mobility, it is necessary to make progress in the direction of increasing its safety.

Transport safety is an essential component of sustainable urban mobility and should be firmly integrated into the mobility planning processes by cities. To stem the road death epidemic, the United Nations have set the target of halving traffic fatalities by 2020. Every year, 1.25 million people are killed in road crashes and up to 50 million are seriously injured. Road crashes kill more people than malaria or tuberculosis and are among the ten leading causes of death globally, and the main cause of death among those aged 15–29 years [1]. Their economic cost is estimated at 2–5% of GDP in many countries. Cyclists represent 8% of all road deaths in the European Union but big disparities exist between countries. Cyclists and pedestrians are generally unprotected and are vulnerable in traffic. As active travel is being encouraged for health, environmental, congestion and other reasons, the safety of walking and cycling in particular must be addressed urgently. Real and perceived safety can have a profound effect on

E. Macioszek and G. Sierpiński (eds.), *Recent Advances in Traffic Engineering for Transport Networks and Systems*, Lecture Notes in Networks and Systems 21,
DOI 10.1007/978-3-319-64084-6_7

modal choice especially in terms of the most sustainable modes of travel – walking and cycling and ability to access public transport. In the ideal case, scenario of urban mobility should be inclusive, safe, resilient and sustainable. Inclusive cities enable citizens to have the freedom to access any opportunity in the city regardless of their social characteristics or economic means. Resilience is the ability for cities to adapt and respond to changes and stresses on social, economic, and environmental systems. Traffic lights that prioritize pedestrian, cyclist, and emergency vehicle traffic puts safety for the most vulnerable on the street first.

Working Group composed by international road safety experts prepared Research Report titled "Zero Road Deaths and Serious Injuries: Leading a Paradigm Shift to a Safe System" which was presented in Paris on 3 October 2016 at a dedicated Road Safety Seminar of The International Transport Forum [2]. This report describes a paradigm shift in road safety policy, according to the principles of a Safe System. A Safe System is based on the premise that road crashes are both predictable and preventable, and that it is possible to move towards zero road deaths and serious injuries. This, however, requires a fundamental rethink of the governance and implementation of road safety policy.

Insufficient attention has been paid to the needs of pedestrians, cyclists and motorcyclists, who together make up 49% of all global road traffic deaths. Making the world's roads safer will not be possible unless the needs of these road users are considered in all approaches to road safety. Traffic planning and safe routes development should be carried out considering characteristics of all categories of road users. It is necessary to take into account the fact that each of these categories can be considered as a separate sub-system of the city transportation system. Its safety and sustainability is ensured by minimizing the number of conflict points that might arise in the interaction of road users. In such systems, possibilities of management (decision-making) based on the information received in real time, have to be available not only to the operators controlling transport, but also to all road users. The objective of such a system is integration of all road users' interaction: people - transport infrastructure - vehicles, with the maximum use of the latest information and control technologies.

2 Safety Issues of the Most Vulnerable Road Users

The number of road traffic deaths has plateaued since 2007 [1]. The plateau in road traffic deaths, set against a 4% increase in global population and 16% increase in motorization, suggests that road safety efforts over the past years have saved lives. However, low-income countries have fatality rates more than double those in high-income countries. Almost half of all deaths on the world's roads are among those with the least protection – motorcyclists (23%), pedestrians (22%) and cyclists (4%). As motorization increases worldwide, walking and cycling have become less common and more dangerous. The traffic mix in many countries means pedestrians and cyclists share the road with high-speed vehicles, forcing them to negotiate dangerous situations and fast moving traffic.

2.1 Technical Measures to Increase Proportion Walking and Cycling in Modal Split

Written by Shannon Bouton, global manager of the McKinsey Center for Business and Environment, and three other senior McKinsey analysts, the report "Urban Mobility at a Tipping Point" finds that "Automobile sales are expected to increase from about 70 million a year in 2010 to 125 million by 2025, with more than half forecasted to be bought in cities" [3]. Authors mention the opinion of automotive analysts who predict that today's 1.2 billion strong global car fleet could double by 2030. "The existing urban infrastructure cannot support such an increase in vehicles on the road", McKinsey warns. Four major technological trends are converging, McKinsey reports. They are "in-vehicle connectivity, electrification, car sharing, and autonomous driving". Walking and cycling are becoming increasingly popular in cities, notes the report, which finds that cities around the world are opening car-free zones to pedestrians and bikers.

Sustainable urban mobility plans and transport strategies aim to reduce mobility demand, shifting to environmentally friendly modes of transport such as walking, cycling and car sharing, increasing the share of public transport and clean vehicles, and the use of state of the art technology for the remaining individual motorized transport. Reducing travel distances and travel demand is therefore an important cornerstone for sustainable urban transport. This imply that the city's inhabitants should not be forced to travel far to fulfill basic needs. Distances should ideally be kept so short that walking or cycling is the attractive/likely mode of transport.

Short distances also improve the possibilities for an enhanced modal shift. This is important to achieve a fair transport system that gives all groups in society equal or high accessibility to important services, independent of car ownership. Reduction of the transportation needs is achieved through the application of sustainable spatial planning polices that support and encourage mixed-use developments. These developments meet the housing, shopping, commercial and employment needs of residents within walking and cycling distance or are accessible by public transport.

Cities around the world are opening car-free zones to pedestrians and bikers. There are many examples of city centers, such as Venice and the medieval city of Mdina, in Malta, that have never allowed motorized vehicles. Mdina is even known as the "Silent City" because of the absence of motor-traffic noise inside the city walls. Pedestrian zones in Japan are called *hokōsha tengoku*, which translates as "pedestrian heaven". Parts of the Calle Florida in Buenos Aires have been restricted to pedestrians since 1913 and the whole street since 1971. More recently, many cities have moved to pedestrianize parts of their city centers. Examples include London, New York, Paris, and Singapore. These efforts often include not only restricting access to cars but also making the streets themselves more attractive to pedestrians by installing better lighting, street signage, and paving materials and adding more greenery.

In addition, many cities are trying to make bicycling safer, easier, and more popular. A study conducted by the Urban Land Institute [4] allows us to understand main regularities in the development of cycling transport. Even in such country, that is oriented to automobile industry, like USA, where almost 70% of population live in private homes, remote 15–30 miles from major transport centers of attraction (jobs, shops, etc.), more than 50% of the residents consider, that cycling and walking accessibility is a top priority when choosing a place of residence, and cycling has become the fastest growing form of passenger transport.

Some cities are experimenting with restricting the entry to the certain streets on weekends; for example, a portion of Central Jakarta is closed on Sunday morning to allow residents to exercise. These efforts can have positive implications for urban mobility. Forty years ago, traffic was terrible in Copenhagen. In 1981 the city developed its first cycling plan and it has been evolving its cycling and mixed-modal goals since 2002. Now transport authorities created a network of bicycle lanes as well as dedicated bicycle highways that reduce traveling time and improve safety. City has a target indicator: 50% of all commutes should be by bike.

It is perfectly feasible for a majority in a metropolis to ride a bike and not travel by car. Not everybody can ride a bike every day, however, which is why the bike should not be seen as a competitor, but rather as complementary to public transport. Especially on the way to and from work there is a lot of potential: in London around 2.5% of all commutes to work are by bike, in Berlin 13%, in Munich 15% and in Copenhagen and Amsterdam a whopping 36 and 37% respectively. Such a high percentage of number of trips to work or education by bicycles in Copenhagen is provided by the fact that the priority strategy of politicians is development of bicycles infrastructure as a way to create more friendly city living condition [5].

Bike sharing has hit the mainstream. In 2015, there were more than a million bikes in bike-sharing programs globally and more than 850 cities had such programs, up from 68 in 2007. One of the most popular counter-argument about cycling are adverse climatic and natural conditions. However, it is a matter of attitude and priority for cycle paths when clearing snow. This is confirmed by the example of Oulu, where a substantial proportion of people commute by bicycle, even when the temperature is below zero in deepest winter. This is ensured by 845 km of routes (4.3 m per inhabitant), 98% of which are maintained throughout winter because main route maintenance priorised over driveways. Routes parallel to driveways are separated with a green lane, which also serves as snow build-up space. There are underpasses in most busy crossings and you can reach every place by bike using cycling routes [6].

London is building 12 "cycle superhighways" – extra-wide lanes dedicated to bicycles. New York expects to have 1,800 miles of bike lanes by 2030. Paris has a bike-sharing network that includes the suburbs and is integrated into the public-transit payment system; São Paolo is doing the same. Delhi is considering proposals to set up separate bike lanes and is providing bike parking near transit stops. Moscow is expanding bike sharing and adding dedicated bike lanes. San Francisco is expanding its network of bicycle lanes; the city's goal is to increase the percentage of all trips taken by bike from 3.4 to 10% by 2018.

Companies in Silicon Valley are piloting electric bikes as means of commuting. At one major technology firm, more than 10% of employees bike or walk to work, and a half of them live within ten miles of the office. A ten-mile commute by car takes about 30 min door to door. By bicycle that would take about 60 min; by e-bike, though, it would be only 35 min, which makes it an attractive option. Getting employees to bike to work is worth real money; each parking space that can be saved on corporate campuses is worth $10,000 to $20,000.

Reduction of conflict points, where the flows of pedestrians, motorized and non-motorized vehicles cross, is the most effective method of increasing safety in transport system. Effective segregation of road traffic is the central requirement of the most campaigners' demands for greater safety, and SkyCycle is one of the most extreme methods of separating bicycles from cars on the agenda. By building a new network of cycle paths high above existing roads and railways, the scheme would create a series of cycle superhighways across London, with specific entry and exit points throughout the city [7]. Dominican architect Richard Morita Castillo offers for solution of transport problem in large cities to build dedicated lanes for cyclists. Innovation lies in placing these lanes over city highway. Such overpasses for cyclists Castillo calls ecobans (like autobahn) or Cycling underground [8]. A similar solution is proposed for Moscow and Kazan [9, 10]. And for Russia it is not only the division of space and diminution the number of conflict points, but also necessary protection for cyclists in difficult weather conditions.

2.2 Organizational Measures of Safe Route's Planning

The quality of the transport system's functioning is largely determined by the quality of management. Operational management in modern conditions is carried out through the intelligent transport systems (ITS) creation. These systems consist of modules for collecting operational information on the status and parameters of the vehicle's operation, analyzing imbalances and searching the causes of their occurrence, as well as developing recommendations for the processes optimization. ITS is increasingly seen as one of the components of solution of the current and future problems in the field of transport. Rational management allows not only increasing the economical performing indicators of the system, but also solving social tasks of the population's transport service improvement, as well as increasing safety of the transport system.

Rational management is based on the "feedback" principle implementation. According to this principle managerial decisions should be corrected in accordance with the identified errors and the system performance deficiencies in the previous period. The purpose of the safety management is reduction of accidents and injuries. This means that the problems of monitoring and conflict situations' detecting, determination of the causes of emergency situations, as well as development of measures to address the causes and reduce the severity of the accidents' consequences are solved.

The main feature of large cities and metropolises is the need to travel long distances both in regular and in one-time trips. Besides, safety and complexity of the route are the determining factors for certain categories of population.

Thus, it is necessary to pay attention to these issues while planning routes. Modern electronic resources (route planners and navigators) allow building a route between two points on the map. The most common of these systems allow selecting the route on the city map specifying one of three types of travel: personal vehicle, public transport and pedestrian route. The quality of the route depends on the degree of elaboration of maps and it is not always possible to obtain adequate information. There are planners designed for individual countries: 2GIS for Russia, jakdojade.pl for Poland, etc. Such planners provide more opportunities to consider individual characteristics while selecting a route. For example, you can select the route for disabled or for cyclists. Despite the large number of applications for path finding, they have one significant drawback: the route is not evaluated from the point of view of its safety.

2.3 Crossroads as a Source of Increased Danger

Pedestrian crossings are safe places for pedestrians to cross the road and where they are given priority. There are various different types of pedestrian crossings, these are: Zebra, Pelican, Puffin, Toucan, Pegasus (also known as Equestrian crossings) [11]. As it is revealed in research [12], almost 30% of pedestrian injuries occur at non-signalized crosswalks. Factors of fatality risks were used to model a binary logistic regression which treated the statistical information provided by the police. Many countries apply signalized pedestrian crossings granting pedestrians the minimal waiting time and allowing the vehicular traffic to continue uninterruptedly if there are no pedestrians in view. The main advantage of signalized pedestrian crossings consists in providing both motorists and pedestrians with a clear perception of the situation at intersection. Instead of solving of a complex logical-physical task, they are offered to primitively react to a light signal with a conditioned reflex. In Europe, where priority is given to pedestrians, there are numerous signalized crosswalks and practically no non-signalized ones across busy four-, and more, lane roads [13].

It is reported [14] that the major reasons for pedestrian noncompliance with traffic signals are the low quality traffic management, traffic volume and long waiting time. In the research [15] it is shown that designing of a signal should take into account the use of neighboring land and the results of traffic survey. Another factor of risk is the conflict between the vehicular and pedestrian flows at left-hand corners [16, 17].

Different models have been proposed for risk assessing and working out of measures to improve the pedestrian safety. Thus, the model proposed in research [18] allows measuring the impact of potential risk factors on pedestrians' intended waiting times during the red-man phase of the traffic lights. In a later research, the author proposes a multivariant method of risk analysis consisting of two hierarchically generalized linear models, characterizing two different facets of unsafe crossing behavior, and uses a Bayesian approach with the data augmentation method to draw statistical inference for the parameters associated with risk exposure. Dependence severity of accidents by vehicle speed is explored in research [19]. The authors took into account age group of pedestrians and vehicle type (sedans and light passenger cars). The authors determined that elderly age group (60 years old and over) faces higher risks of serious injury and fatality than both the child age group (12 years old and under) and the mid-age group (13–59 years old). The authors suppose that the findings should be included in designing specifications of pedestrian detection system in the near future.

3 Results and Discussion

3.1 Using the Haddon Matrix to Improve Traffic Safety

The Haddon Matrix, developed by William Haddon in 1970, looks at factors related to personal attributes, vector or agent attributes and environmental attributes; before, during and after an injury or death. By utilizing this framework, one can then think about evaluating the relative importance of different factors and design interventions. In fact, it is a set of preventive measures to improve road safety. In this context, primary prevention refers to interventions before the event, to prevent it completely; secondary prevention involves lessening the extent of injury given that an event occurs (e.g., wearing a seatbelt will not prevent the collision but may lessen its effects). Tertiary prevention limits the subsequent difficulties a person encounters given his injuries (rehabilitation, etc.). You could also add primordial prevention, which would refer to underlying social change that would alter the circumstances in which the accident occurred. For example, encouraging public transportation would reduce roads' overloads and likely prevent some accidents.

The adoption of measures to improve road safety, as a rule, is based on the analysis of statistical data on offenses and traffic accidents. This analysis allows you to adjust the traffic patterns, make changes to infrastructure and make recommendations to manage traffic [20]. Official statistics of the State Traffic Safety Inspectorate (STSI) of Naberezhnye Chelny city as well as the data from the official website of the STSI of Russian Federation is used as initial information. Preventive measures to improve the road safety include identification of violations that can be divided into two categories: violations that increase the possibility of an accident (drink–driving, leaving the oncoming traffic) and violations that increase the severity of accident consequences (disuse of child restraints and seat-belts). You could also add primordial prevention which would refer to underlying social change that would alter the circumstances in which the accident occurred. Such events may be, for example, planning safe routes, which eliminate or minimize risks, and include measures that will reduce the severity of the consequences in the occurrence of risk situations.

3.2 Proposed Algorithm of the Safest Route's Selection

In constructing the routes of urban mobility it is necessary to choose the safest possible routes, taking into account characteristics of the participant movements. The categorization of the route can be performed in the presence of different factors, complicating traffic conditions. Depending on the type of move routes in the city can be road, pedestrian, cycling and combined. The combined route usually includes areas where the movement takes place by public transport, and the movement between the start (end) point and public transport, on foot or by bike. These areas are evaluated by different criteria groups because for different categories of the population priorities may be different. For example, the selection by the criterion "fastest route" may just not realize for people with disabilities, and in some cases this route may be unsafe. Therefore, we propose a methodology a multi-criteria evaluation of the safety of the route. The integrated circuit evaluation algorithm is shown in Fig. 1.

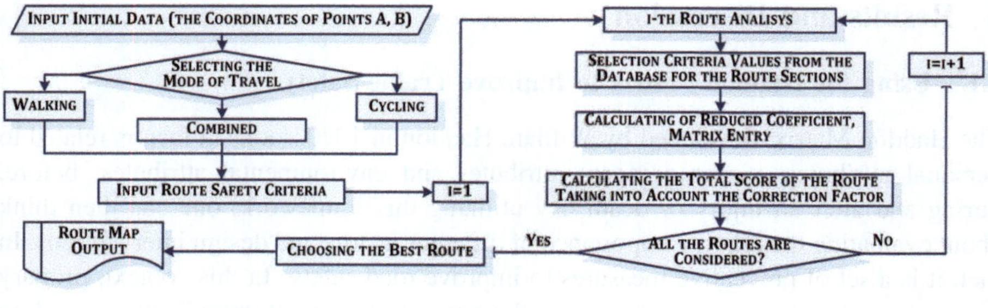

Fig. 1. The integrated algorithm of the route safety assessment

In the first stage there were the factors, which determine the category of the route complexity (Table 1) for pedestrians and cyclists. Since the significance of the factor for different categories of traffic participants is determined by their features to determine the weight factors have been used for the category pedestrians and cyclists by age and state of health (Table 2). The evaluation of the route can be done with the help of a complex indicator, which is calculated as the weighted average:

$$K = \sum_{i=1}^{n} K_i \cdot \alpha_i \tag{1}$$

where:
K_i - the value of the i-th parameter,
α_i - weight of the index.

In order to summarize dissimilar indices in the formula their reduced values are calculated.

Table 1. Factors influencing the route's safety.

Factors	Significant factor for	
	Pedestrians	Cyclists
The length of the route	+	+
The length of sections with elevation changes	+	+
The number of pedestrian crossings	+	+
The number of unregulated intersections	+	+
The number of lanes of the highway at the place of pedestrian crossings	+	+
The number of ramps	+	+
The number of underground/raised pedestrian crossings	+	+
The length of bikeways on the route	–	+
The presence of places of concentration of road accidents on the route	+	–

Table 2. Types of pedestrians and cyclists.

Children	Youth, Middle-aged people, Retired	Disabled
Preschoolers	Without health problems	Doesn't require special conditions for movement
Students	With problems	Needing special conditions

Factors that determine the safety of the route can be both objective (e.g., terrain, presence of unregulated intersections, etc.), and subjective due to the features and physical condition of the road user (age, health condition, etc.). Adequacy of the assessment will depend on the correctness of the selected factors and their combined inclusion. For example, the same route can be safer in the daylight than in the dark, in the summer than in the winter, etc. Possible routes are evaluated on the base of this information. To do this a matrix of the given route options is constructed and then the overall routes' performance indicators are calculated. The value of the route safety indicator is calculated with provision for correction factors that depend on the physical condition and characteristics of the user. A significant part of all urban movements is a mixed routes that include parts of the route where people go on foot, by bicycle or by public transport. That is why when assessing the route along with the safety of walking and cycling areas the number of transfers on the route, because the safety of such interchange nodes can be a determining factor.

3.3 An Example of Route Planning: Case Study of Naberezhnye Chelny

The route between gymnasium "A golden hill" and Organ Hall can be considered as an example. This relatively small length route (from 1.2 to 1.4 km) has its own peculiarities. Part of the route passes through the yards, the most long part passes along Moskovskiy pr. and the small part passes along perpendicular streets: Yashlek prospect and Kamal boulevard. There are three lanes in each driving direction of Moskovskiy prospect, that is one of the main longitudinal streets of the city. There are only two traffic lights at the beginning and at the end of the considered area and there are two pedestrian underpasses near the bus stops. Besides that, there is a non-signalized crosswalk (Zebra) that due to the specifics of the traffic organization in this area is unsafe. As the Fig. 2 shows, Yandex Maps (YM) and Google Maps (GM) lay almost the same routes that differ from each other by the way of passing through the yards. However, both of them propose to use unsafe non-signalized Zebra crosswalk though there are two pedestrian underpasses in the considered area.

The second feature of this route is that it is too long for elderly people (seniors), physically disabled people and pedestrians with small children. It is possible to solve this problem by using public transport, i.e. combined route. YM system proposes to use a tram that follows from the point "Rembyttekhnika". This route is comfortable because you can cross the road using the signalized crosswalk. However, there is another alternative route that is not considered by YM system: if you use the pedestrian underpass that is located just before the tram stop, you can travel one stop less and,

Fig. 2. The route laid by Google (a) and Yandex (b) Maps

what is more, you can use tram of any route. After selecting the "Less walking" option GM system proposes two variants. The first is to take the bus following through a turn in the direction, opposite to the destination and then to cross the road and to use public transport in the necessary direction (Fig. 3a). The second is to travel to the right direction directly after crossing the road (Fig. 3b).

At the same time specification of these maps is insufficient to construct an optimal route. Moreover, many residents of the city would have used a completely different route, which is not indicated neither in YM system, nor in GM system. A more adequate route is provided by the 2GIS search system (Fig. 4).

The given example shows that development of tools for comparison and evaluation of possible routes from the point of view of safety and acceptability for different population groups, is a good addition to existing route planners.

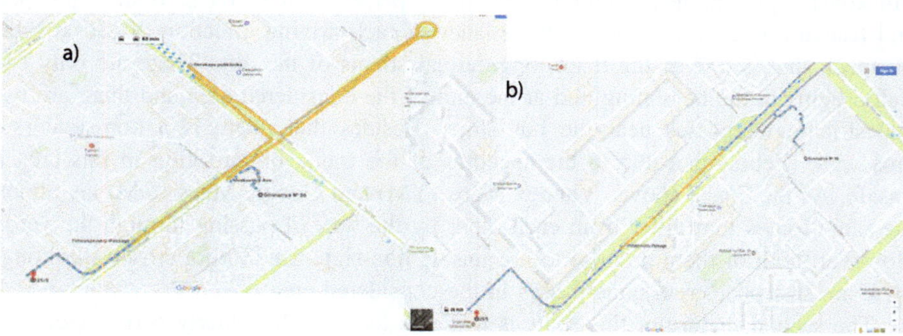

Fig. 3. The route laid with "Less walking" option

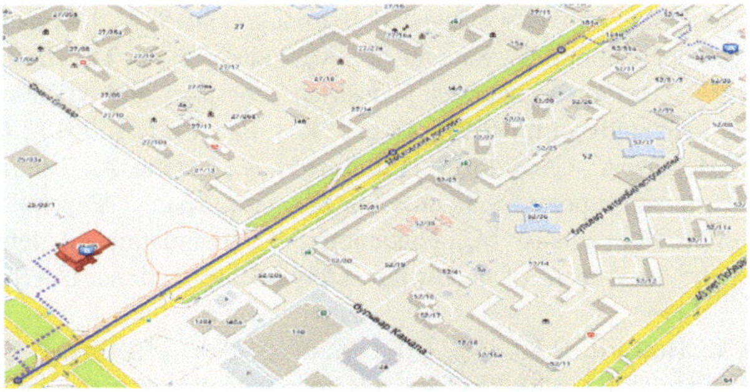

Fig. 4. The route laid by 2GIS

4 Conclusions

Safety of pedestrians and cyclists is an actual problem. Its relevance increases due to the growth of motorization and its negative effects. To improve the effectiveness of measures aimed at reducing the risks of road accidents the special tool is needed. It should allow responding quickly to changes in the transport system and to assess the effectiveness of the measures taken. Proposed system for the route's safety assessment and evaluation of the effectiveness of measures aimed at reducing the risks of dangerous situations will allow making justified managerial decisions.

References

1. World Health Organization. http://www.who.int/violence_injury_prevention/road_safety_status/2015/GSRR2015_Summary_EN_final.pdf
2. International Transport Forum. http://www.itf-oecd.org/road-safety-seminar-safe-system
3. McKinsey. http://transact.org/mckinsey-report-urban-mobility-tipping-point/
4. Urban Land Institute: Active Transportation and Real Estate: The Next Frontier. Urban Land Institute, Washington (2016)
5. Bredal, F.: The case of Copenhagen. In: Roethig, M., Efimenko, D. (eds.) Changing Urban Traffic and the Role of Bicycles. Russian and International Experiences, pp. 24–28. Friedrich-Ebert-Stiftung, Moscow (2014)
6. Tahkola, P.: The case of Oulu. In: Roethig, M., Efimenko, D. (eds.) Changing Urban Traffic and the Role of Bicycles. Russian and International Experiences, pp. 29–43. Friedrich-Ebert-Stiftung, Moscow (2014)
7. Construction Chat. http://www.constructionchat.co.uk/articles/skycycle-london-bike-network/
8. 5dimensions. http://05dimensions.wordpress.com/2012/10/22/bici-metro-an-alternative-project-to-sky-cycles-is-available-for-global-implementation
9. Smart News. http://smartnews.ru/regions/kazan/14990.html
10. Velomesto. https://velomesto.ru/magazine/in-russia/na-vdnh-mozhet-poyavitsya-velopoliten/
11. World Driving. http://www.driving-school-beckenham.co.uk/pedestriancros.sigs.html
12. Olszewski, P., Szagała, P., Wolański, M., Zielińska, A.: Pedestrian fatality risk in accidents at unsignalized zebra crosswalks in Poland. Accid. Anal. Prev. **84**, 83–91 (2015)

13. The National Archives. http://www.legislation.gov.uk/uksi/1997/2400/made
14. Brosseau, M., Zangenehpour, S., Saunier, N., Miranda-Moreno, L.: The impact of waiting time and other factors on dangerous pedestrian crossings and violations at signalized intersections: a case study in montreal. Transp. Res. Part F **21**, 159–172 (2013)
15. Di Stasi, L.L., Megias, A., Cándido, A., Maldonado, A., Catena, A.: The influence of traffic signal solutions on self-reported road-crossing behavior. Span. J. Psychol. **17**, 1–7 (2014)
16. Alhajyaseen, W.K.M., Asano, M., Nakamura, H.: Estimation of left-turning vehicle maneuvers for the assessment of pedestrian safety at intersections. IATSS Res. **36**, 66–74 (2012)
17. Quistberg, D.A., Howard, E.J., Ebel, B.E., Moudon, A.V., Saelens, B.E., Hurvitz, P.M., Curtin, J.E., Rivara, F.P.: Multilevel models for evaluating the risk of pedestrian-motor vehicle collisions at intersections and mid-blocks. Accid. Anal. Prev. **84**, 99–111 (2015)
18. Li, B.: A model of pedestrians intended waiting times for street crossings at signalized intersections. Transp. Res. Part B **51**, 17–28 (2013)
19. Oikawa, S., Matsui, Y., Doi, T., Sakurai, T.: Relation between vehicle travel velocity and pedestrian injury risk in different age groups for the design of a pedestrian detection system. Saf. Sci. **82**, 361–367 (2016)
20. Makarova, I., Khabibullin, R., Belyaev, A., Mavrin, V.: Urban transport system management in the context of region sustainable development strategy. Trans. Probl. **8**(2), 107–111 (2013)

Impact of the Factors Determining the Natural Stack Effect on the Safety Conditions in a Road Tunnel

Aleksander Król[1] and Małgorzata Król[2(✉)]

[1] Faculty of Transport, Silesian University of Technology, Katowice, Poland
aleksander.krol@polsl.pl

[2] Faculty of Energy and Environmental Engineering, Silesian University of Technology, Gliwice, Poland
gosia.krol@polsl.pl

Abstract. Impact of weather conditions and natural stack effect on the operation of the mechanical ventilation system in a road tunnel is considered in the study. The work presents the measurements conducted in an opened road tunnel. Then, with the use of Ansys Fluent software the numerical model was built and validated. The numerical analyses were extended to other tunnel inclinations and temperature differences between tunnel portals. Taguchi method of experiment design was used to diminish the number of required simulations and to analyse the results.

Keywords: Road tunnel · Fire ventilation · Natural stack effect · Ansys fluent

1 Introduction

The development of the road infrastructure is associated with the construction of road tunnels. There are either urban tunnels, which aimed to relieve urban centers or non-urban tunnels, whose task stems from topographical conditions. All the tunnels must be equipped with ventilation systems.

A ventilation system in a road tunnels should provide an acceptable level of communication pollutants and a sufficient visibility [1]. Moreover, in a case of fire, it should before all support conditions for safe evacuation of the people and reinforce the work of rescue teams. Such ventilation system in a road tunnel can be implemented as natural or mechanical. The selection of a relevant ventilation system depends on the length of the tunnel, the road traffic intensity and the traffic mode (unidirectional or bidirectional) [2, 3].

Additionally the operation of ventilation system is affected by natural stack effect. This phenomenon involves air movement in whole tunnel volume without any artificial forcing. Its intensity depends on the tunnel configuration, especially on tunnel inclination and tunnel orientation, the local topography and the weather conditions [4, 5]. Depending on the combination of these factors natural stack effect is a fairly stable or may be variable.

E. Macioszek and G. Sierpiński (eds.), *Recent Advances in Traffic Engineering for Transport Networks and Systems*, Lecture Notes in Networks and Systems 21,
DOI 10.1007/978-3-319-64084-6_8

The measurement of natural stack effect was one of components of the research conducted in an opened road tunnel. The tunnel is located in the S69 road between towns of Żywiec and Zwardoń in Southern Poland. Its popular name is Laliki Tunnel. It is a typical example of non-urban tunnel, which supports the convenient pass under the hill.

There are the measurements of the intensity of natural stack effect, its impact on fire ventilation operation and the numerical model presented in this work. The numerical model was applied to study on the factors influencing the intensity of the natural stack effect: tunnel inclination, wind speed and temperature difference on both portals (Fig. 1).

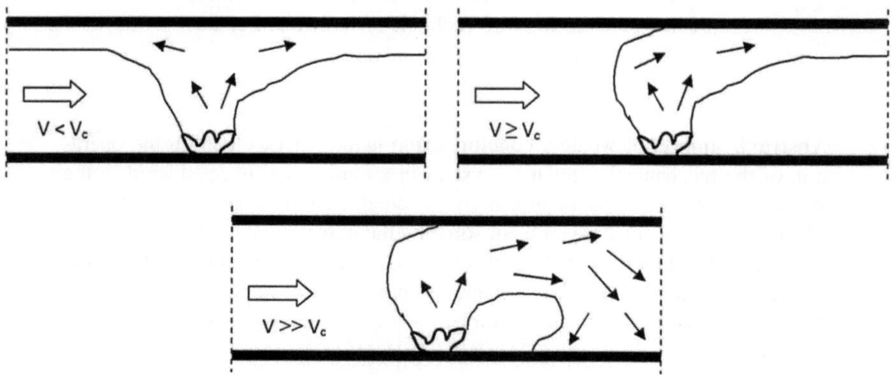

Fig. 1. Formation of a smoke layer for different relations between real air velocity and required critical velocity, V – real airflow velocity, V_c – critical velocity

2 The Airflow in Tunnels

The aim of fans operation in a road tunnel is forcing the air movement in determined direction. The velocity of airflow is of great importance for proper operation of fire ventilation system. During the fire outbreak the velocity of airflow should high enough to ensure the flow of smoke and heat in required direction. However, this velocity must not be too high to avoid disturbance of natural stratification of smoke and air layers. There is a critical velocity, which should be determined for each tunnel to protect against the flow of smoke in the opposite direction than the required.

It is expected, that if the air velocity is not lower than the critical velocity, the hot smoke produced by the fire will move according the direction of the longitudinal ventilation operation and will not move into opposite direction [6].

The phenomenon of smoke movement into the opposite direction than the direction of operation of the longitudinal ventilation is called backlayering. This phenomenon is strongly undesirable because it causes smoke to contaminate the entire tunnel despite the operation of the longitudinal ventilation [7]. Critical velocity can be determined according to the following expressions [2]:

$$V_c = K_1 K_g \left(\frac{gHQ}{\rho C_p A T_f} \right)^{\frac{1}{3}} \quad (1)$$

$$T_f = \left(\frac{Q}{\rho C_p A V_c} \right) + T \quad (2)$$

where:

V_c - critical velocity [m/s],
K_1 - 0.606 (Froude number),
K_g - inclination factor (depends on tunnel inclination),
g - gravitational acceleration [m/s^2],
H - tunnel height [m],
Q - heat release rate (HRR) [kW],
ρ - air den sity [kg/m^3],
C_p - air specific heat [kJ/kg K],
A - area of tunnel cross-section [m^2],
T_f - fire gases temperature [K],
T - air temperature [K].

The above presented system of equations need to be solved iteratively. The inclination factor (K_g) should be taken into account for tunnels with slope. It can be determined with the use of graph included in NFPA Standard [2] or calculated analytically as follows [8]:

$$K_g = \left(1 + 0.0374 |\min(grade, 0)|^{0.8} \right) \quad (3)$$

The inclination of a tunnel causes additional airflow related to the natural stack effect. It is assumed, that if the inclination is greater than 1–2%, this fact should be taken into account when designing the ventilation system and determining the critical velocity [9, 10]. The impact of natural stack effect can be of great importance in case of inclined long tunnels. If the difference of temperature between portals appears (e.g. due to insolation at one mountainside) it can cause spontaneous air movement in the tunnel. Similar effect will occur in a case of fire: the equilibrium of air layers will be disrupted and hot air and smoke will start to move up. Additional contribution comes from the wind and strongly depends on its direction and velocity [11, 12].

Provisions of the Directive 2004/54/WE include the impact of natural stack effect on the operation of a ventilation system and limit the allowed inclination of a tunnel to 5%, unless other circumstances make it impossible [1].

3 Description of the Examined Tunnel

The tunnel is located at S69 road between towns Żywiec and Zwardoń. The length of the tunnel is 678 m, its width is 11.9 m and height is 6.55, the ceiling is vault shaped. The tunnel is constantly inclined, the slope is 4%. It is slightly curved (Fig. 2). The northern portal is placed at the height of 669 m a.s.l., the southern portal is placed at the

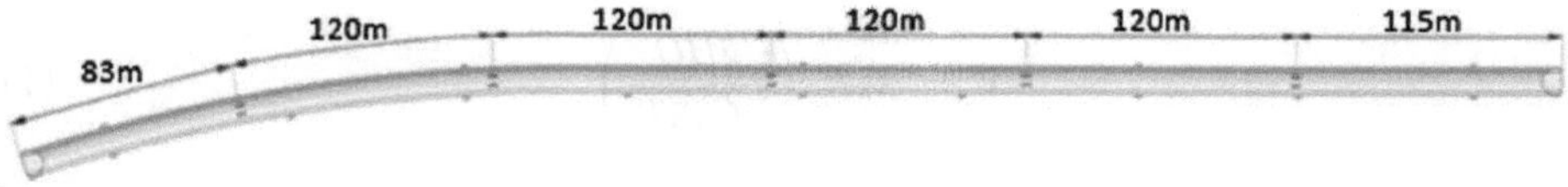

Fig. 2. The overall sketch of the Laliki tunnel

height of 642 m a.s.l. The tunnel goes under a low hill, which however, affect the diversity of meteorological conditions at the both portals.

The tunnel is divided functionally into five section, they are numbered from northern portal. There are 2 fans with thrust of 810 N each in every section. All the fans are reversible, but their normal operation direction is S → N. The possibility of reversible work is very important due to emergency operation pattern in a case of fire. The operation of the fans is intended not to spread up the smoke over entire tunnel. Basically, if it is possible, the smoke should be forced to move towards the nearest portal, rather than be pushed over the major part of the tunnel.

4 Measurements of the Airflows

The research involves the measurements of the velocity of airflows at different modes of fans operation. There were sets of Sensor Company anemometers used for this purpose (Fig. 3).

Fig. 3. Sensor Company thermal anemometer

Each anemometer consists of a pair of spherical sensors of 3 mm diameter made of enameled copper wire. Both sensors are heated, the temperature difference between the sensor exposed to the air flow and the second one is a measure of air velocity. The measuring range is from 0.05 to 5 m/s, the accuracy is ±0.02 m/s. Each anemometer sent the averaged value every 30 s. A single measurement lasted 3 min and the average of 10 such readings was regarded as the result. The measurements were repeated at some other localizations for different fans operation modes.

The airflows velocity due to natural stack effect was measured directly too. The hand-held anemometer which was here used was an element of weather station Kestrel 4500. During whole the research the weather conditions were recorded at both portals of the tunnel: weather station Kestrel 4500 at the southern portal and ultrasonic anemometer WindMaster Pro Gill Instruments at northern portal. All measurements had to take place within one day, taking advantage of the opportunity to shut down the tunnel traffic.

5 The Results

The atmospheric pressure was 942 hPa and was almost unchanged during the whole day. The temperature was also almost stable and varied between 7 and 8 °C. In one of the experiments the air velocity was measured at different heights and in two modes of fans work (normal and reverse). A stand pole with a set of anemometers was placed at the symmetry axis of the tunnel, at a distance of 45 m from N portal. One fan section S5 was turned on. Figure 4 shows the results of the measurements: distribution of air velocity at tunnel axis for normal and reverse work mode of fans is compared to the critical velocity. As can be seen, in normal mode of fans work (S → N) the critical velocity is provided at the lower part of tunnel, it ensures the smoke removal. At the upper part of the tunnel the air velocity is smaller than the critical velocity, what prevents against mixing the upper layer of smoke with the fresh air below.

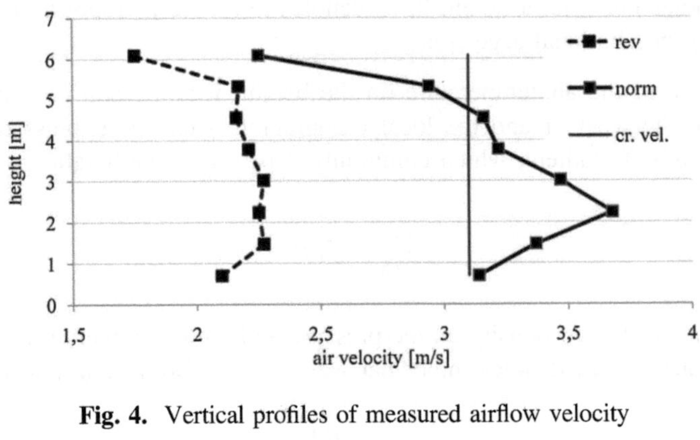

Fig. 4. Vertical profiles of measured airflow velocity

In the reverse mode the air velocity at all measurement points is lower than the critical velocity. Because the fans are claimed to be fully reversible, it is caused mainly by the strong natural stack effect in the examined tunnel. This is not a threat in the case of Laliki tunnel, because there is a possibility of turning on another fans section.

The velocity of airflow caused by the natural stack effect (all fans were turned off) was measured directly at the height of 2 m in the middle of the tunnel length at a few points along the width. The measured values ranged from 1 to 1.5 m/s and were very unsteady. In the same time the wind was variable (as within the whole day): at the northern portal it blew from directions from W to NE with varying speed 0–3 m/s, at the southern portal the wind direction was from SSW to W and its speed varied in the range 0–2.5 m/s. The described above results proved the existence of natural caused air flow from south to north, the effect is significant and is able to impact on the efficiency of ventilation system. These measurements were qualitatively confirmed by the historical records of air velocity inside the tunnel – the ultrasonic sensor placed under the tunnel ceiling was recording the air velocity continuously. The obtained data showed the existence of natural air flow from south to north of velocity varying between 0.5 and 3 m/s.

6 The Phenomenon of Natural Stack Effect in an Inclined Tunnel

Intensity of the natural stack effect is difficult to predict [13]. It is caused by the complexity of the process and by the its dependence on many interacting factors. The most important factors, which mutually contribute to formation of natural airflow in an inclined tunnel and determine its intensity are:

- drop of the atmospheric pressure with the increasing height, described by the approximate formula:

$$p = p_0 - \rho g(h - h_0) \tag{4}$$

where p denotes pressure at height h, p_0 denotes reference pressure at height h_0, ρ is density and g gravitational acceleration.

- dependence of the air temperature on the height, it is a result of actual weather, history of the weather and the local topography; it can be expressed as the real temperature (*T*) gradient, which commonly depends on the height:

$$\frac{\Delta T}{\Delta h} = f(h) \tag{5}$$

- dependence of the air density on the pressure and on the temperature, according to the ideal gas law (μ denotes molecular weight, k_B is Boltzmann constant):

$$\rho = \frac{\mu p}{k_B T} \tag{6}$$

- the air properties – if a volume of air moves vertically, it expands or compresses (depending on the movement direction); because the thermal conductivity of air is rather low, the process is usually considered as adiabatic [14], hence the temperature is changing according to the formula (κ is adiabatic exponent):

$$\frac{T}{T_0} = \left(\frac{p}{p_0}\right)^{\frac{\kappa-1}{\kappa}} \tag{7}$$

- the wind influence – the natural stack effect depends on wind speed and wind direction,
- the airflow in the tunnel causes pressure drop between the portals; this pressure drop is expressed by the Darcy-Weisbach law (λ is a constant depending on the flow details, D is equivalent diameter of the tunnel):

$$\frac{\Delta p}{\Delta l} = \frac{\lambda}{D}\frac{\rho v^2}{2} \tag{8}$$

As can be seen all these phenomena interact with each other and they strongly depend on the initial and boundary conditions. Thus the convenient way to estimate velocity distribution in the tunnel is to build a numerical model and to solve it.

Such a simplified numerical model of Laliki road tunnel was built with the use of Ansys Workbench platform and solved by Ansys Fluent software. The model did not take into account the tunnel curvature, so it was possible to apply Ansys Fluent feature – symmetry plane, what allowed to significant decrease of computational complexity of the problem. A short summary of details of the numerical model is presented in Table 1. The tunnel being analyzed is rather long and should be regarded as a pipe, thus the *k*-omega turbulence model was chosen due the best reproducing the flows in the vicinity of walls.

Table 1. Numerical model summary.

Feature		Value
Turbulence model		*k*-omega SST
Fluid material		Air (ideal gas)
Operating pressure		94,200 Pa
Gravitational acceleration		9.81 m/s^2
Solver		Pressure based
Pressure/velocity coupling		Simple
Under-relaxation factors	Pressure	0.1
	Momentum	0.1
	Energy	0.95
	Others	Default

Because the wind velocity during the measurement was varying at both portals taking values from the range 0–2 m/s the numerical analyses were done for three wind velocities: 0, 1, 2.5 m/s. The ambient temperature was set to 8 °C as was recorded in the reality. As results the averaged velocity magnitude and averaged longitudinal component of velocity at the cross-section at the middle of the tunnel length was obtained (Table 2).

Table 2. Averaged air velocities in the middle of the tunnel.

Averaged air velocity	Wind velocity at southern portal [m/s]		
	0	1	2.5
Velocity magnitude [m/s]	0.62	0.55	1.38
Longitudinal component of velocity [m/s]	0.08	0.27	1.27

The results are presented also in Fig. 5, the graphs show the distributions of velocity magnitude and longitudinal component of velocity at the height of 2 m at the middle of the tunnel length, perpendicularly to its axis.

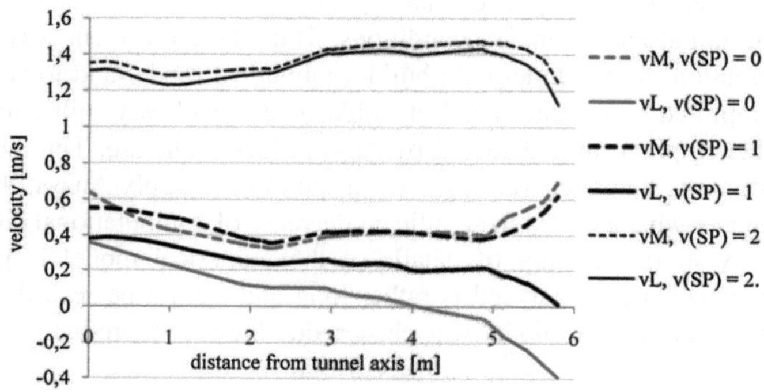

Fig. 5. Velocity magnitude (vM) and longitudinal component of velocity (vL) at the height of 2 in the middle of the tunnel for 3 wind velocities at southern portal v(SP)

As can be seen for low wind velocities the longitudinal component of velocity (which is responsible for straight air movement) is very low and its contribution to the total air flow is negligible. In such situation the chaotic airflows prevail. Moreover, in the area close to the tunnel walls the air flows towards southern portal appears. For higher wind velocities the longitudinal component of air velocity determines the total air movement.

The calculated numerically values of the air velocity due to natural stack effect fit to the averaged measurements. The exact comparison is impossible because the measured values were very unsteady.

Table 3. Parameters sets for extended analysis of the intensity of natural stack effect.

Parameter		Value		
		1	2	3
Tunnel inclination [%]	TS	0	3	6
Wind velocity [m/s]	WS	0	3	6
Temperature difference [°C]	DT	0	2	4

After validation of the model the scope of numerical research was extended to examine the intensity of natural stack effect for other possible tunnel inclinations, other temperature differences between tunnel portals and different wind velocities. These input parameters are shown in Table 3.

As can be seen, the full factorial analyses, even for such scant selection of parameters would require up to 27 independent simulations. Because of great number of possible parameters combination the Taguchi method can be applied to diminish the number of simulations to perform [15]. If there are 3 parameters which can take 3 values, the appropriate experiment scheme is described by the L9 orthogonal array. According to this scheme there are only 9 numerical simulations needed. The parameters and the results for each simulation are shown in Table 4. The longitudinal component of the air velocity averaged for the cross-section in the middle of the tunnel length is regarded as the result.

Table 4. The L9 orthogonal array and the simulation results.

No	TS level	WS	DT	TS [%] value	WS [m/s]	DT [°C]	VL
1	1	1	1	0	0	0	−0.01
2	1	2	**2**	0	3	2	**1.84**
3	1	3	3	0	6	4	5.47
4	2	1	3	3	0	4	0.30
5	2	2	1	3	3	0	1.85
6	2	3	**2**	3	6	2	**5.54**
7	3	1	**2**	6	0	2	**0.15**
8	3	2	3	6	3	4	2.10
9	3	3	1	6	6	0	5.52

To find out the relative impact of each parameter the results should be now averaged for each level for each parameter as in example for parameter DT and level 2 (2 ° C):

$$R_{DT2} = \frac{\left(v_{avg1} + v_{avg2} + v_{avg3}\right)}{3} \tag{8}$$

The final analysis of relative impact of parameters on the stack effect is shown in Table 5.

Table 5. Analysis of the relative impact of parameters.

	TS	WS	DT
1	2.43	0.15	2.45
2	2.56	1.93	2.51
3	2.84	5.51	2.62
min	2.43	0.15	2.45
max	2.84	5.51	2.62
Δ	0.41	5.36	0.17

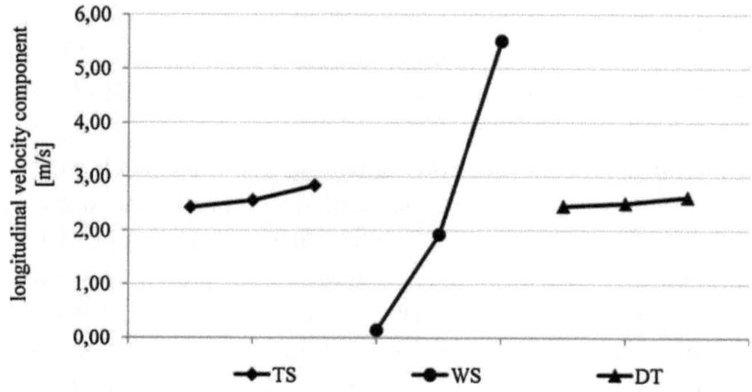

Fig. 6. The influence of the parameters on the natural stack effect

These dependences can be presented also as a graph (Fig. 6), which is clearly showing how the particular parameter influences the natural stack effect.

The prevailing role of the wind is proved without a doubt. It is especially important when designing a non-urban tunnel: such tunnels are often located in mountain areas, where winds occur according to a local pattern. The impact of the other factor is much less significant, however they can cause perceptible air flow in the absence of the wind.

7 Conclusions

The presented analyses showed clearly the great impact of natural stack effect on conditions in a road tunnel and then on the efficiency of ventilation system. The natural stack effect caused by the tunnel inclination and the wind influence force the intense air flow.

It seems to be very important while designing a tunnel, especially an inclined one to estimate the intensity of natural stack effect. The CFD methods can be very useful for this task. The historical weather data and the local wind rose should be taken into account. The obtained results should be treated as an important premise for following stages of design.

Acknowledgments. The authors thank General Directorate for National Roads and Motorways (Katowice Division) and Voivodeship Headquarters of State Fire Service in Katowice for the opportunity of the research and the management support. We thank also the Smay Company and the Ardor Company for the technical support.

References

1. European Parliament: Directive 2004/54/EC of the European Parliament and of the Council of 29 April 2004 on Minimum Safety Requirements for Tunnels in the Trans-European Road Network. European Parliament, Brussels (2004)
2. National Fire Protection Association: NFPA 502: Standard for Road Tunnels, Bridges and Other Limited Access Highways. NFPA, Quincy (2011)
3. Ministerstwo Transportu i Gospodarki Morskiej: Rozporządzenie Ministra Transportu i Gospodarki Morskiej z dnia 30 maja 2000 r. w sprawie warunków technicznych, jakim powinny odpowiadać drogowe obiekty inżynierskie i ich usytuowanie. (Dz. U. Nr 63, poz. 735 z póź. zm.). Ministerstwo Transportu i Gospodarki Morskiej, Warsaw (2000)
4. Zhong, W., Fan, C.G., Ji, J., Yang, J.P.: Influence of longitudinal wind on natural ventilation with vertical shaft in a road tunnel fire. Int. J. Heat Mass Transf. **57**, 671–678 (2013)
5. Tanaka, F., Kawabata, N., Ura, F.: Effects of a transverse external wind on natural ventilation during fires in shallow urban road tunnels with roof openings. Fire Saf. J. **79**, 20–36 (2016)
6. Klote, J.H., Milke, J.A., Turnbull, P.G., Kashef, A., Ferreira, M.J.: Handbook of Smoke Control Engineering. Ashrae, Atlanta (2012)
7. Tang, F., Li, L.J., Mei, F.Z., Dong, M.S.: Thermal smoke back-layering flow length with ceiling extraction at upstream side of fire source in a longitudinal ventilated tunnel. Appl. Therm. Eng. **106**, 125–130 (2016)
8. Musto, M., Rotondo, G.: Numerical comparison of performance between traditional and alternative jet fans in tiled tunnel in emergency ventilation. Tunn. Undergr. Space Technol. **42**, 52–58 (2014)
9. Chow, W.K., Gao, Y., Zhao, J.H., Dang, J.F., Chow, C.L., Miao, L.: Smoke movement in tilted tunnel fires with longitudinal ventilation. Fire Saf. J. **75**, 14–22 (2015)
10. Lin, P., Lo, S.M., Li, T.: Numerical study on the impact of gradient on semi-transverse smoke control system in tunnel. Tunn. Undergr. Space Technol. **44**, 68–79 (2014)
11. Sztarbała, E., Sztarbała, G.: Systemy Wentylacji Pożarowej Tuneli Drogowych. Chłodnictwo & Klimatyzacja **5**, 78–82 (2015)
12. Sztarbała, E., Sztarbała, G.: System Mechanicznej Wentylacji Wzdłużnej Tuneli Drogowych w Warunkach Pożaru. Chłodnictwo & Klimatyzacja **8**, 91–95 (2015)
13. MaoHua, Z., CongLing, S., Li, H., JieHong, S., Chang, L., XiangLiang, T.: Smoke development in full-scale sloped long and large curved tunnel fires under natural ventilation. Appl. Therm. Eng. **108**, 857–865 (2016)
14. Hartmann, D.L.: Global Physical Climatology, 2nd edn. Elsevier, Amsterdam (2016)
15. Roy, R.K.: Design of Experiment Using the Taguchi Approach. John Wiley & Sons, New York (2001)

Studying the Behaviour of Pedestrians and Drivers Within Pedestrian Crossings

Marcin Budzyński[1(✉)], Kazimierz Jamroz[1], Łukasz Jeliński[1], and Mariusz Kieć[2]

[1] Faculty of Civil and Environmental Engineering, Gdańsk University of Technology, Gdańsk, Poland
{mbudz,kjamroz,lukjelin}@pg.gda.pl
[2] Faculty of Civil Engineering, Kraków University of Technology, Kraków, Poland
mkiec@poczta.onet.pl

Abstract. Every third road accident in Poland involves a pedestrian as a participant or, most of the time, a casualty. Pedestrian accidents are usually the result of complex situations and the outcome of a number of factors related to driver and pedestrian behaviour and road infrastructure. Safety depends largely on how well the traffic condition is perceived and on visibility in traffic. The relations between pedestrians and motorists and the resulting hazards should researched to ensure that pedestrians are provided with the right conditions for walking.

Keywords: Pedestrian safety · Driver behavior · Pedestrian behavior

1 Introduction

The distance to be covered by a road user is the decisive factor when road users choose how they will travel. More than 75% of walks are under 1 km with only 5% of the trips above 2 km. Transport behaviour studies in Poland show [1] that walking represents 20–25% of all trips in cities. To compare, in Paris the share of walking is app. 36% and 32% in London [2]. People walk primarily for shopping, public transport stops, school or workplace, for recreation and tourism, public utility places, etc.

With no clear and understandable rules for pedestrian priority, drivers' behaviour may pose a risk to pedestrians. This includes drivers driving carelessly across pedestrian crossings and pedestrians carelessly stepping onto the road, crossing the road where it is not allowed, etc. Lack of adequate pedestrian infrastructure on the road (refuge islands, signage) and along the road (pavements) is another problem. This is particularly true for city outskirts, sections of transit roads and sections of roads outside built-up areas where pedestrian traffic may be quite intense. In addition, some of the existing pedestrian facilities do not meet the standards of traffic and safety (e.g. technical condition, location of pedestrian crossings related to sight distance, public transport stops, deficient pedestrian areas).

E. Macioszek and G. Sierpiński (eds.), *Recent Advances in Traffic Engineering for Transport Networks and Systems*, Lecture Notes in Networks and Systems 21,
DOI 10.1007/978-3-319-64084-6_9

The paper presents the results of analyses of methodologies for systematic studies of pedestrian behaviour and pedestrian-driver relations (for the Secretariat of the National Road Safety Council) [3].

2 Pedestrian Hazards in Poland

In 2014 there were 1.116 pedestrian road deaths on Polish roads (34% of all fatalities), 8.397 pedestrians were injured, of which 3.009 were serious injuries. While the trend in pedestrian casualties in Poland is successively decreasing (Fig. 1), the percentage share of pedestrian deaths in all road deaths remains high which in 2014 was 34%. The demographic fatality rate continues to be very high for pedestrian fatality accidents. Last year it was 29 people per 1 million population.

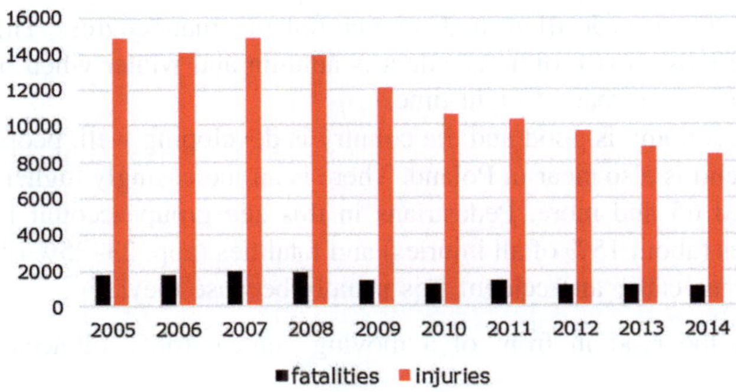

Fig. 1. Distribution of pedestrian injury and fatality in Poland between 2005–2014

The standard of Poland's road network differs from region to region. As a result, the safety of road users differs as well. The risk of being involved in an accident is particularly high for pedestrians. The degree of the risk can be measured using the demographic rate of serious pedestrian accidents (pedestrian fatality and serious injury per 100.000 population). The analysis was applied to the regions and showed that (Fig. 2):

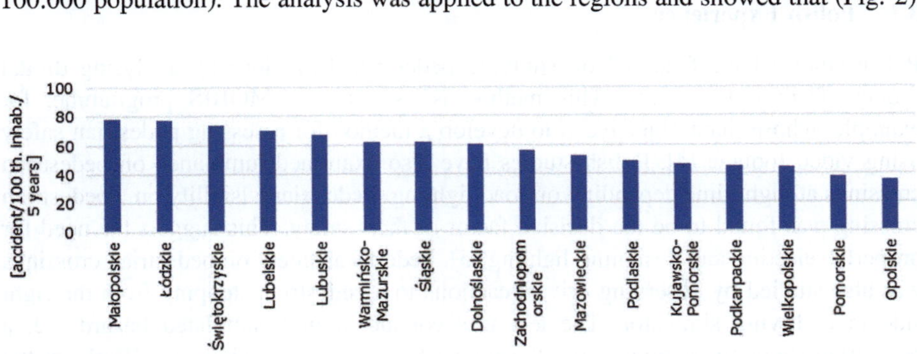

Fig. 2. Ranking of societal risk measured with the demographic indicator of serious pedestrian road accidents

- the highest risk of serious pedestrian accidents was in the regions of Małopolskie and Łódzkie,
- the lowest risk was in the regions of Opolskie and Pomorskie.

Pedestrian traffic is usually concentrated in urban areas which typically generate traffic (public utility buildings, workplace, education, entertainment). This is why it is important to design this space in a pedestrian-friendly manner as much as possible.

Over the analysed period more than 90% of injuries in pedestrian accidents and more than 65% of fatalities happened in built-up areas. Pedestrian accidents usually happen on the road - nearly 60% of all pedestrian accidents. Pedestrian crossings are the next element of road infrastructure where pedestrians are at risk of an accident - more than 30%, and app. 4% on the pavement. The share of injuries is similar while fatalities are as follows: road –76%, pedestrian crossings –19%, pavement –1%.

An overwhelming majority of pedestrian accidents are caused by drivers of passenger cars. The most frequent causes include speed, failure to give way on a pedestrian crossing to a pedestrian and non-compliant manoeuvres. The time when pedestrians are most at risk of an accident is autumn and winter when pedestrians are difficult to spot on the road at night-time.

When the economy is good and the country is developing well, people tend to live longer. This trend is also clear in Poland. There is an increasingly higher proportion of road users aged 65 and more. Pedestrians in this age group account for the highest share of injuries (about 15% of all injuries) and fatalities (app. 23–25% of all fatalities). When pedestrians cause an accident it is usually because they:

- stepped on the road in front of a moving vehicle (58% of accidents, 50% of fatalities),
- stepped on the road from behind a vehicle (10%),
- crossed the road where it is not designed for crossing (11%).

3 Experience of Pedestrian Behaviour and Pedestrian - Driver Relation

3.1 Polish Experience

Polish studies have focussed on studying pedestrian behaviour by analysing digital footage from test cameras. This method is used in the MOBIS programme, for example, whose main objective is to develop a method for assessing pedestrian safety using video footage [4]. Polish studies have also examined luminance on pedestrian crossings at night time depending on road lighting. Pedestrian visibility on a pedestrian crossing was found to be the decisive factor of their safety. This suggest the need for properly selecting and designing lighting [5]. Pedestrian safety on pedestrian crossings was also studied by observing driver reactions to a pedestrian stepping from the right side in a driving simulator. The test was conducted in a simulated hazard, i.e. a pedestrian suddenly stepping onto the road while a car is moving. [6]. Work on the textbook "Pedestrian Safety" [7] included pilot in-the-field tests to collect more reliable data. The tests included observations of conflicts and dangerous situations. Another approach to studying pedestrian-driver behaviour was applied in Warsaw for assessing the city's pedestrian transport system. Pedestrians were surveyed to understand: road safety, technical condition of pavements, ease of access to public transport stops, ease of movement (such as lack of obstacles, i.e. cars parked on pavements) [8].

3.2 International Experience

In the United States pedestrian traffic is usually studied using surveys. US scientists have developed several types of surveys with the most popular ones being: a walkability checklist and principles of pedestrian design. When the checklist is used, pedestrians are handed a base list with pre-defined problems and the answers are on a scale from 1 to 6 [9]. A pedestrian - driver survey was conducted in Maine using cameras to record the survey. The analysis looked at the probability of an accident on pedestrian crossings for different parameters such as driver speed, width, a marked or unmarked pedestrian crossing, signalised or unsignalised crossing, with barriers or without them [10, 11].

In Australia a survey was conducted to establish how familiar road users are with traffic regulations on the priority of road users in different situations [12]. Another research method applied in Australia looked at the behaviour of children on main and local roads. The study focussed in particular on where they stop, whether they look around, analyse how vehicles move and how children cross the street [13]. In another study surveys were used to collect the basic parameters such as the respondent's age, gender and education. The analysis also looked at the destination and knowledge of the regulations by asking questions about road users' priorities in different situations [14]. A study in New Zealand was designed to analyse time lost by pedestrians on crossings with traffic lights. This was measured by observers who would randomly choose a pedestrian approaching a crossing and record the time when they stopped and crossed the street [15].

A study in France aimed to analyse the behaviour and mutual pedestrian-driver relations. When the results were analysed it was found that when a pedestrian looks at an approaching vehicle, the vehicle is more likely to stop by more than 10% [16]. A survey in the United Kingdom checked three potentially dangerous behaviours: crossing a dual carriageway, crossing on red on a Pelican light controlled crossing and crossing a very busy road between parked cars [17]. The method used in Spain involved automatic pedestrian detection on crossings using cameras. The parameters measured included: waiting time, pedestrian and vehicle speed, vehicle type and pedestrian gap acceptance. In addition, manual checks were made of gender, age, number of people or pedestrian baggage [18]. In Germany behaviour in road traffic at signalised junctions was studied recording the duration of signal phases, time for a pedestrian to cross the whole crossing or part of it (when the road was divided by a dividing lane or refuge), waiting time, pedestrian speed, "type of pedestrian behaviour" and driver-pedestrian conflicts [19]. In the Netherlands studies showed that excessive vehicle speed is the biggest problem for pedestrians, especially for people aged >65. This is due to difficulty in assessing the speed of oncoming vehicles. It was also established that young people frequently cross on red [20]. In Sweden road user behaviour was recorded with cameras [21, 22]. In Belgium a study was made of whether pedestrians respect the red light on signalised crossings [23]: 21% of the pedestrians did not comply. In Finland cameras were used to study vehicle and pedestrian speed, traffic volume and driver behaviour. Based on this an estimate was made of the number of drivers who stop before the crossing, slow down or do not change the speed [24]. Austria's method was divided into four stages: surveys by telephone and on test sites. Drivers were surveyed on test sites and asked to identify dangerous or non-compliant situations [25].

The SARTRE 4 project is an element of the European Road Safety Observatory (ERSO). In the first three editions of the project only drivers were surveyed; in the fourth edition the target group was extended to include motorcyclists, cyclists, pedestrians and public transport passengers. Thirty three questions asked about the behaviour and sense of road safety. The answers in Poland showed that 22% of respondents felt very safe on the roads, 46% felt partially safe, 32% thought the roads were dangerous [26].

As we can see from Polish and international research, there is a need for site tests to observe pedestrian and driver behaviour within pedestrian crossings. Equally, surveys are important to understand road users' impressions and attitudes to specific solutions and regulations and to learn what problems they have observed. There is a need for studying real situations on pedestrian crossings, depending on where they are located and arranged, pedestrian volumes and pedestrian profiles. In the case of drivers it is important to observe how drivers react depending on pedestrian behaviour. This helps with assessing the effectiveness of the solutions and improving pedestrian safety where it is particularly dangerous, i.e. on pedestrian crossings. The majority of the countries known to study pedestrian behaviour and driver-pedestrian relations, do not conduct regular studies and only focus on selected aspects. Systematic research in Poland is needed especially due to the very high risk for pedestrians.

4 Pilot Tests in Poland

The scope of the pilot projects was limited to two regions: Pomorskie - it has an average level of pedestrian risk (the second region with the lowest societal risk in relation to the number of fatal and serious injury accidents vs. the population), Małopolskie - it has a very high level of pedestrian risk (the worst region for societal risk in relation to the number of fatal and serious injury accidents vs. the population). The areas vary in terms of the development, road cross-section and location of pedestrian crossings (Fig. 3).

One of the site parameters was the speed of vehicles approaching the pedestrian crossing. A set of digital cameras were used for top quality recording. The location of the cameras made sure that they covered the entire crossing as well as the approach (app. 100 m.) with test point cross-sections spaced every 10 m. In addition, vehicles in free-flow and selected pedestrian and driver behaviours were taken into account.

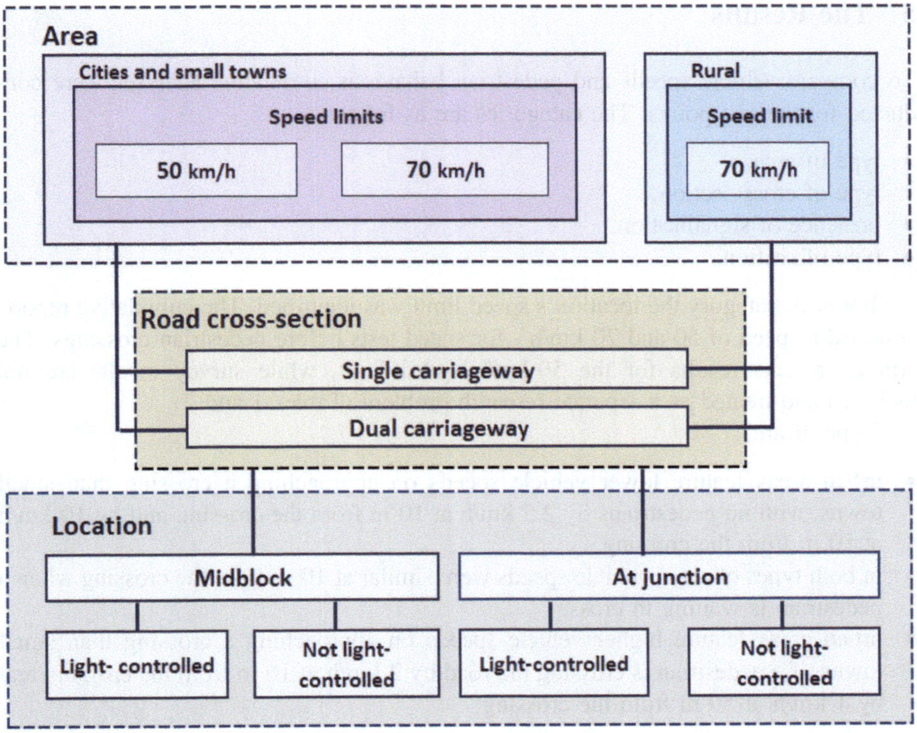

Fig. 3. Diagram of test points selection. Source: [3]

The following situations were identified for analysing vehicle speeds on pedestrian crossings:

- no pedestrian,
- a pedestrian waiting before the crossing,
- a pedestrian on the pedestrian crossing.

Pedestrian behaviour was studied looking at the following characteristics: distance between the pedestrian and road while waiting to cross, whether the pedestrian looked at the traffic before and during crossing, pedestrian age structure and gender. In addition, pedestrians were surveyed after they had crossed the road as well as drivers who parked near selected test points.

In both regions 69 test points were designated, 1890 surveys were conducted with pedestrians, which is on average 30 surveys per one test point. There were also 543 surveys with drivers.

The problems which emerged in the pilot and which need to be addressed in national tests is that some pedestrian crossings were used by a very small number of pedestrians throughout the day (less than 20) which is particularly frequent in places outside built-up areas and in small towns.

5 The Results

To compare vehicle speeds and pedestrian behaviour cumulative analyses were conducted for the test points. The categories are as follows:

- type of area,
- type of cross-section,
- presence of signalisation,
- type of section.

For each category the location's speed limit was identified. The cumulative reports assumed a speed of 50 and 70 km/h - for speed tests before pedestrian crossings. The article presents results for the 50 km/h speed limit, while survey results are not included and treated as a separate research problem (Tables 1 and 2).

Type of area:

- urban areas feature lower vehicle speeds on approaching a crossing than small towns, with no pedestrians by 2.5 km/h at 10 m from the crossing and by 10 km/h at 50 m from the crossing,
- in both types of areas vehicle speeds were similar at 10 m from the crossing when a pedestrian is waiting to cross,
- urban areas feature higher vehicle speeds on approaching a crossing than small towns, if a pedestrian is crossing the road by 2 km/h at 10 m from the crossing and by 1 km/h at 50 m from the crossing,

Table 1. Characteristics of vehicle speeds within pedestrian crossings - type of area (Source: [3])

Situation	Average speed *V* [km/h] at *L* [m] from the crossing								Speed percentiles			Number of vehicles	All number of vehicles
	0	10	20	30	40	50	60	70	V.15	V.85	V.95		
Type of area													
Urban area													
No pedestrian	44.8	44.6	45.3	46.3	43.7	43.5	45.1	38.8	31.5	52.6	58.9	2908	3489
Pedestrians waiting	43.4	43.4	42.4	44.2	37.4	38.1	38.9	42.1	29.7	51.2	57.3	1319	1714
Pedestrians crossing	0.0	13.2	21.0	27.4	30.2	33.8	33.8	39.0	8.0	26.0	31.9	959	1294
All	36.6	38.4	40.0	42.5	39.6	40.5	42.1	39.6	24.1	54.0	61.2	5186	6497
Small town													
No pedestrian	47.0	47.0	46.6	47.6	55.4	53.1	64.6		37.1	56.4	63.6	1223	1345
Pedestrians waiting	43.7	43.7	40.7	35.3	30.1	37.0	55.0		44.2	55.0	58.2	87	87
Pedestrians crossing	0.0	11.3	17.2	22.0	27.8	34.7	50.5		20.7	27.4	30.4	76	76
All	44.3	44.8	44.7	45.4	52.1	50.5	64.5		33.3	56.2	63.1	1386	1508

Table 2. Characteristics of vehicle speeds within pedestrian crossings-type of cross-section (Source: [3])

Situation	Average speed V [km/h] at L [m] from the crossing							Speed percentiles			Number of vehicles	All number of vehicles
	0	10	20	30	40	50	60	V.15	V.85	V.95		
Type of cross-section												
1 × 2 cross-section, no refuge												
No pedestrian	49.3	49.4	49.3	52.2	53.8	52.2	58.8	35.3	54.1	60.0	1193	1413
Pedestrians waiting	26.1	26.1	24.3	28.7	30.2	35.6		38.4	49.5	53.1	197	240
Pedestrians crossing	0.0	11.4	19.6	27.8	31.4	33.6	46.7	15.5	26.2	30.1	182	227
All	40.7	41.7	42.7	45.4	47.2	49.9	58.7	36.3	50.3	56.7	1527	1880
1 × 2 cross-section, with refuge												
No pedestrian	27.8	27.2	30.2	30.4	38.5	41.4	28.7	37.1	56.4	63.6	966	1255
Pedestrians waiting	42.0	42.0	39.0	33.4	28.4	36.5	17.5	44.2	55.0	58.2	86	113
Pedestrians crossing	0.0	10.9	21.1	20.1	24.7	33.6	27.1	20.7	27.4	30.4	269	376
All	23.1	24.9	28.9	29.4	35.6	39.9	28.3	33.3	56.2	63.1	1321	1744
1 × 4 cross-section												
No pedestrian	57.3	57.3	58.8	62.3	60.5	62.5	64.6	37.1	56.4	63.6	353	396
Pedestrians waiting	41.2	41.5	37.7	45.0	40.6	39.2	55.0	44.2	55.0	58.2	310	403
Pedestrians crossing	0.0	12.5	22.3	31.9	34.6	34.8	50.5	20.7	27.4	30.4	192	259
All	38.8	41.7	43.2	49.4	51.1	53.0	64.5	33.3	56.2	63.1	865	1058
2 × 2 and 2 × 3 cross-section												
No pedestrian	50.9	50.7	49.9	49.0	44.9	41.4	64.6	37.1	56.4	63.6	1560	2184
Pedestrians waiting	48.7	48.6	48.7	47.8	40.8	38.2	55.0	44.2	55.0	58.2	812	1177
Pedestrians crossing	0.0	15.5	20.2	26.3	29.2	33.7	50.5	20.7	27.4	30.4	382	515
All	44.0	45.2	45.4	45.5	41.1	39.1	64.5	33.3	56.2	63.1	2754	3876

- in both types of areas vehicle speeds were slightly lower at 10 m from the crossing if a pedestrian is waiting to cross as opposed to when there are no pedestrians and speeds were substantially lower when pedestrians were crossing the road.

Type of street cross-section:

- the lowest speed on approaching a crossing with no pedestrians at 10 m from the crossing was recorded for the 1 × 2 cross-section with refuge (by 30 km/h less than for 1 × 4, and 23.5 km/h less than for dual carriageways),

- the lowest speed on approaching a crossing if a pedestrian is waiting to cross was recorded for the 1 × 2 cross-section with no refuge (by 16 km/h less than for 1 × 4 and 1 × 2 with refuge, 22 km/h less than for dual carriageways),
- the lowest speed on approaching a crossing if a pedestrian is crossing the road was recorded for the 1 × 2 cross-section with refuge but the differences between the cross-sections are very small.

For all types of cross-sections, vehicle speed when pedestrians wait to cross is lower than when there are no pedestrians (by 16 km/h for 1 × 2, by 16 km/h for 1 × 4, by 2 km/h for 2 × 2 and 2 × 3).

The lowest speed on approaching a crossing with no pedestrians waiting, at 10 m from the crossing was recorded for the 1 × 2 cross-section with a refuge island (30 km/h less than for 1 × 4, 23 km/h less than for dual carriageways (Fig. 4). The lowest speed on approaching a crossing with pedestrians waiting was recoded for 1 × 2 cross-section with no refuge (16 km/h less than for 1 × 4, 22 km/h less than for dual carriageways). The lowest speed on approaching a crossing being crossed by a pedestrian was recorded for 1 × 2 with a refuge but the differences between the cross-sections are quite small.

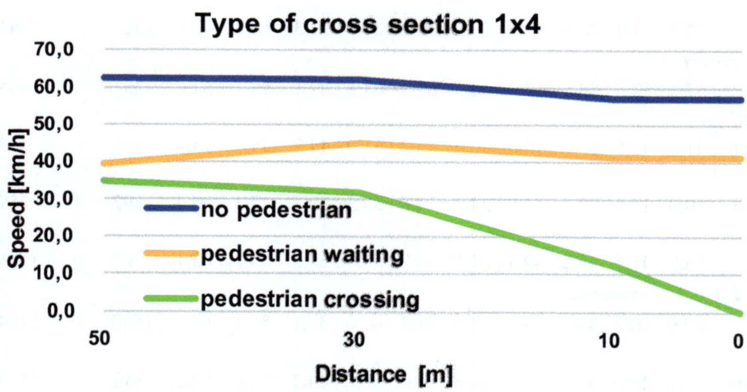

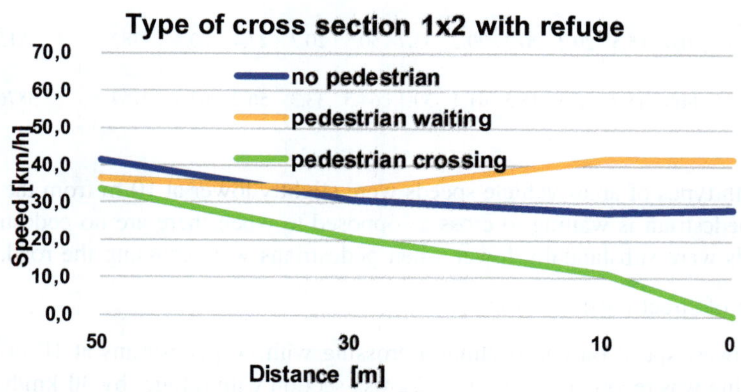

Fig. 4. Characteristics of vehicle speeds on approaching pedestrian crossings for selected cross-section. Source: [3]

6 Conclusion

Walking plays a crucial role in the transport system. This is true of small towns and villages with very little public transport and of big cities where walking is often used to move around the city. Walking is also part of many people's everyday lives (especially children, school youth, older people, people who do not have a car). Having said this, pedestrians are the most vulnerable road users and most at risk of death in road accidents, representing more than 30% of all road accident fatalities in Poland.

The paper presents the results of analyses of methodologies for systematic studies of pedestrian behaviour and pedestrian-driver relations. The effects of the location of the site, type of cross-section and other selected parameters on pedestrian and driver behaviour are demonstrated.

The relations between walking and driving and the hazards generated should be further researched to ensure that pedestrians can use roads safely.

The pedestrian-driver relation and the behaviour must be monitored on an on-going basis and pedestrian crossings must be inspected for safety. This will help improve pedestrian safety effectively.

References

1. Development Office of Gdansk: Comprehensive Study of Traffic in Gdansk. Development Office of Gdansk, Gdansk (2009)
2. Transport for London: Making London a Walkable City. The Walking Plan for London. Pedestrian Comfort Level Guidance. Transport for London, London (2005)
3. Jamroz, K., Budzyński, M., Gaca, S., Kieć, M.: Methodologies for Systematic Studies of Pedestrian Behaviour and Pedestrian - Driver Relations. National Road Safety Council, Warsaw (2015)
4. Olszewski, P., Buttler, I., Czajewski, W., Dąbkowski, P., Kraśkiewicz, C., Szagała, P., Zielińska, A.: Pedestrian safety assessment with video analysis. Transp. Res. Procedia **14**, 2044–2053 (2016)
5. Tomczuk, P.: Pedestrian safety—luminance measurements at a pedestrian crossing. Prace Naukowe Politechniki Warszawskiej. Seria Transp. **80**, 117–135 (2011)
6. Guzek, M., Jurecki, R.S., Karendał, M., Lozia, Z., Zdanowicz, P.: Research drivers' reaction to the pedestrian coming from the right, performed in a driving simulator. Buses **6**, 1–11 (2010)
7. Jamroz, K., Gaca, S., Michalski, L. et al.: Pedestrian safety. In: Handbook for Organizers Pedestrian Traffic. National Road Safety Council, Warsaw (2014)
8. Transeko: Diagnosis and Rating Pedestrian System in Warsaw. Transeko, Warsaw (2010)
9. Partnership for a Walkable America: Walkability checklist. http://www.walkableamerica.org/checklist-walkability.pdf
10. Gårder, P.E.: The impact of speed and other variables on pedestrian safety in Maine. Accid. Anal. Prev. **36**, 533–542 (2004)
11. Schroeder, B.J.: A Behavior-Based Methodology for Evaluating Pedestrian-Vehicle Interaction at Crosswalks. North Carolina State University, Raleigh (2008)

12. Hatfield, J., Fernandes, R., Job, R.F.S., Smith, K.: Misunderstanding of right-of-way rules at various pedestrian crossing types: observational study and survey. Accid. Anal. Prev. **39**(4), 833–842 (2007)
13. Drummond, A.E., Ozanne-Smith, J.: The Behaviour and Crash Involvement Risk of Child Pedestrians and Bicyclists: A Traffic Exposure Study. Monash University, Wellington (1991)
14. Fildes, B., Lee, S., Kenny, D., Foddy, W.: Survey of Older Road Users. Behavioral & Travel Issues. Monash University, Wellington (1994)
15. Thomas, J., Rive, G., Frith, B., Charlton, S.: Pedestrian-Vehicle Accident Review: Human Factors Behavioural Assessment. Opus, Gracefiels (2011)
16. Guéguen, N., Meineri, S., Eyssartier, C.: A pedestrian's stare and drivers' stopping behavior: a field experiment at the pedestrian crossing. Saf. Sci. **75**, 87–89 (2015)
17. Evans, D., Norman, P.: Understanding pedestrians' road crossing decisions: an application of the theory of planned behaviour. Health Educ. Res. **13**, 481–489 (1998)
18. Alonso, I.P., Llorca, D.F., Sotelo, M.A., Bergasa, L.M., Revenga, P., Nuevo, J., Ocana, M., Garrido, M.: Combination of feature extraction methods for SVM pedestrian detection. IEEE Trans. Intell. Transp. Syst. **8**(2), 292–307 (2007)
19. Ni, Y.: Pedestrian Safety at Urban Signalised Intersections. Technische Universitat, Darmstadt (2010)
20. Hummel, T.: Dutch Pedestrian Safety Research Review. Federal Highway Administration, Georgetown Pike (1999)
21. Várhelyi, A.: Dynamic Speed Adaptation Based on Information Technology: A Theoretical Backround. University of Lund, Lund (1997)
22. Johansson, C.: Towards a Method to Improve Road Safety for Pedestrians and Cyclists. Especially in Child Pedestrian Environments. Lulea University of Technology, Lulea (2001)
23. Diependale, K.: Do Pedestrians Comply with Traffic Lights in Belgian Cities? http://www.ictct.org/wpcontent/uploads/Diependaele1.pdf
24. Pasanen, E.: Traffic Safety at Pedestrian Zebra Crossings. Ministry of Transport and Communications, Helsinki (2007)
25. Berger, W.J., Hössinger, R.: Car Drivers' Attitudes Towards Non-Signalised Zebra Crossings and Their Impact on Behaviour. University of Natural Resources and Life Sciences, Vienna (2010)
26. Papadimitriou, E., Theofilatos, A., Yannis, G.: Patterns of pedestrian attitudes, perceptions and behaviour in Europe. Saf. Sci. **53**, 114–122 (2013)

Failure Evaluation of the Level Crossing Protection System Based on Fault Tree Analysis

Waldemar Nowakowski[1(✉)], Tomasz Ciszewski[1], Jakub Młyńczak[2], and Zbigniew Łukasik[1]

[1] Faculty of Transport and Electrical, Kazimierz Pulaski University of Technology and Humanities in Radom, Radom, Poland
{w.nowakowski,t.ciszewski,z.lukasik}@uthrad.pl
[2] Faculty of Transport, Silesian University of Technology, Katowice, Poland
jakub.mlynczak@polsl.pl

Abstract. An intersection where railway line crosses a road at the same level is called level crossing. The development of the automotive industry has resulted in an increase in the number of vehicles. Thus the traffic on the level crossings has been raised and this, in turn, has resulted in an increase requirements in terms of level crossings safety. For this purpose, the Level Crossing Protection Systems (LCPS), that help to protect these road users, are built. LCPS systems, due to the need to ensure the efficient operating of the railway, should be developed taking into account the detailed analyses. It is necessary to include in these analyses the growing needs and expectations of users in terms of system functionality. However, the requirements for ensuring the reliability and safety of systems should be taken into consideration above all. One of the methods to identify and analyse the factors which may cause adverse events is the Fault Tree Analysis (FTA). The authors of this article using the FTA method carried out a qualitative analysis of the safety of B class level crossings. Traffic safety related requirements for LCPS system were described. Then this analysis was used to identify top and intermediate events and build necessary fault trees. Conducted analysis may be helpful in the process of designing new LCPS systems.

Keywords: Level crossing protection systems · Fault tree analysis · Safety assessment reliability

1 Introduction

On the railway, the level crossings are particularly important places, because of the need of ensuring the safety. Accidents at level crossings cause substantial material damages and multiple fatalities. Such significant consequences have resulted from the weight of trains and long braking distance. Therefore, there is a need to carry out detailed safety analysis in order to limit these damages. According to CENELEC standards all railway applications are obliged to conduct reliability and safety analysis [1–4]. One of the methods to identify and analyse the factors which may cause adverse

E. Macioszek and G. Sierpiński (eds.), *Recent Advances in Traffic Engineering for Transport Networks and Systems*, Lecture Notes in Networks and Systems 21,
DOI 10.1007/978-3-319-64084-6_10

events is the Fault Tree Analysis (FTA) [5–8]. The authors of this article using the FTA method carried out a qualitative analysis of the safety of B class level crossings. Traffic safety related requirements for LCPS system were described. Then this analysis was used to identify events and build necessary fault trees.

2 Level Crossing Protection Systems

2.1 Principle of Operation

Level crossing protection systems are used to alert users of roads about approaching trains, and thus are an important component of the railway infrastructure affecting the railway safety as well as the traffic safety [9, 10]. Warning is accomplished by means of road signals and acoustic signal generators (buzzer or bell). In addition, the level crossing can be equipped with barriers and warning shields. The principle of operation of the B class level crossings (Fig. 1) is discussed later in the article. Warnings are triggered when the train is approaching to the level crossing, as a result of the train detection at the activation points (WS1.1, WS2.2). Wheel sensors and induction loops are usually used as the activation points. In the case of control Sect. 1 the wheel sensor WS1.1 is used, while control Sect. 2 uses the wheel sensor WS2.2. Information about the approaching train is transmitted to the central control unit, which initiates the warning. Then the road signals S1, S2 and acoustic signals are activated, and provide the appropriate signal on the level crossing warning shield for the selected direction. Another step is the closure of the barriers.

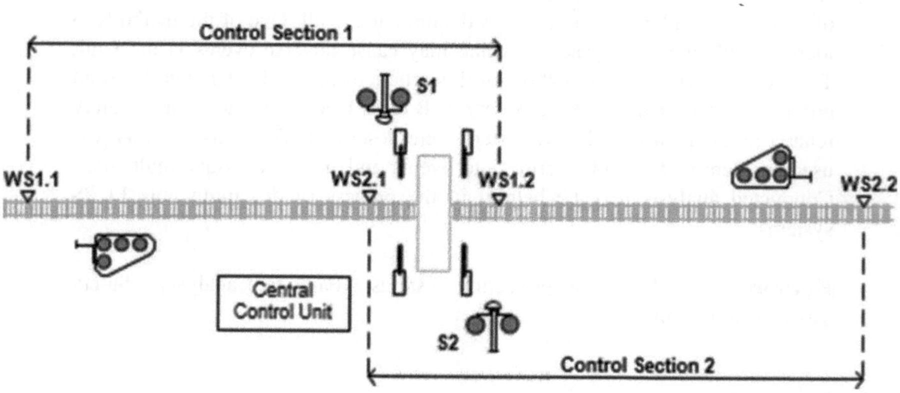

Fig. 1. LCPS equipment and its impact zone on the level crossing (single-track railway line)

When a train is getting closer to an intersection the deactivation points (WS2.1, WS1.2) causes the switching off an acoustic signal. If the passing zone is left by the train the warning shield are disabled and the barriers starts opening. After opening the barriers road signals are deactivated. The manual control panel allows the manual closing and opening of the level crossing in special cases [11].

In modern LCPS the remote diagnostic and automatic collection of information about operation of level crossings are often used. These features are available by the application of the PLCs (Programmable Logic Controller) in central control unit. Concept of modern central control unit based on the reliability structure "2 of 2". Central control unit additionally supervise the peripheral groups for train detection, roadside signals and barriers. System faults are detected by comparing the information between the two independent channels [12, 13].

2.2 Safety Requirements

LCPS depending on its application and the related level crossing class must fulfil different requirements. Nevertheless, common to this group of systems are the following requirements:

- turning on the warning when the train is approaching to the level crossing, considering the time needed to arrive to the level crossing,
- turning off the warning immediately when the train leaves the level crossing,
- moving away from the level crossing should not cause warnings (especially in the case of single-track railway line where trains travelling in both directions and share the same track),
- entering the next train into the control section should provide warning,
- changing the direction of the train, after leaving the control section should cause warning shutdown,
- in the case of using barriers, closing them should take place with a certain delay after the switching on the warning lights and acoustic signals,
- the barriers shall be equipped with barrier boom lights,
- the opening of the barriers by unauthorized persons should be impossible.

LCPS system in case of any failure needs to ensure railway traffic safety. Therefore, all faults have been divided into two categories: "I category", which directly threaten the safety of railway traffic, "II category" do not threaten safety.

In case of the I category fault occurrence the system initializes proceedings that lead to limit the speed of the train. Reduced speed allows safely stop the train in case of obstacles on the level crossing. This category include:

- failures of the activation points of the wheel sensors,
- too low voltage of power batteries,
- failures detected by maintenance/diagnostic systems (such as the lack of communication between the controllers),
- road signalling damages,
- damages of barriers.

Another group are the II category faults, which do not endanger the safety of railway traffic, and thus do not enforce the speed limit of trains. This category include:

- failures of the deactivation points of the wheel sensors,
- loss of power supply voltage,

- the lack of charging voltage at the output of chargers,
- failure of the bulb in the level crossing warning shield,
- opening the cabinets door of the central control unit.

In the case of the occurrence of this type of failures the LCPS system alerts the operating staff.

As already mentioned, the main requirement for LCPS is to ensure high reliability and safety of railway traffic. Therefore, the analysis of failures should be carried out having regard to commonly acknowledged methodologies [14, 15]. An example of such method might be a Fault Tree Analysis (FTA), which the authors of the publication used for analyse LCPS system failures.

3 Fault Tree Analysis

Fault Tree Analysis (FTA) is a standardised, advanced, deductive technique which is used to identify and analyse the factors that may be the cause of the adverse events [5, 7, 8, 16]. FTA is thus a graphical model of the cause-and-effect relationships. Fault trees are used for qualitative analysis, when we try to recognize the risky situation, and also for quantitative analysis, which allows to determine the probabilities of sequences of events. FTA enables modelling the course of failure, and then its analysis using logical trees. FTA allows to identify the causes which effects is defined as an uncertain event or risk. These causes affixed to the fault tree are interrelated and may be referred to as: specific failures (for example, the equipment failures), human errors, environmental conditions, other events which may lead to failure.

The logical relationships of the event in the fault tree are shown by gates. Each gate is connected with one output event and one or more input events. The events should be labelled, so that it can be easy to identify the relationship between fragments of fault tree. The AND type and OR type gates are the most commonly used in fault trees.

A combination of events described by AND gate means that output event occurs only if all of the input events occur. In terms of system reliability this implies system failure only in case of all components failure. In case of two independent events A and B connected into inputs of AND gate, the probability of occurrence of the output event P_{AB} is given by the Eq. (1):

$$P_{AB} = P(A \cap B) = P(A) * P(B) \tag{1}$$

Whereas the OR gate is used output event occurs if any of the input events occurs. In terms of system reliability, this gate implies system failure if any of system components fail. Output event probability P_{AB} for two events, in case of OR gate, is given by the formula (2):

$$P_{AB} = P(A \cup B) = P(A) + P(B) - P(A \cap B) \tag{2}$$

If the inputs events A and B are mutually exclusive events then $P(A \cap B) \approx 0$ and it can be assumed, that output event probability is:

$$P_{AB} \approx P(A \cup B) = P(A) + P(B) \tag{3}$$

Fault tree analysis usually includes the following steps [17, 18]:

1. Definition of the problem and the boundary conditions:
   - identification of the top event, which allows to find critical paths leading to the occurrence of this event,
   - definition of the boundary conditions for the analysis.
2. Construction of the fault tree:
   - determination of hierarchical structure of fault tree and identification of sub-events,
   - fault tree construction by linking the events using logic gates,
   - identification of the basic events leading up to the top event.
3. Finding minimal cut sets/or path sets in combined basic event leading up to the top event.
4. Qualitative analysis of the fault tree:
   - analysis of the results for the designation of the importance of the basic events, by finding the relative contribution to the top event probability from a basic event.
5. Quantitative analysis of the fault tree:
   - estimation probability of the occurrence of the basic events,
   - calculation probability of the occurrence of the top event.

Additionally sensitivity analysis, which allows to find these basic events, where a relatively small change in their probability will lead to relatively large change in the probability of top event can be carried out.

3.1 Fault Tree Construction

Fault tree analysis displays the logical relationships in the chain of events leading to a failure in level crossing protection system. One of the strengths of fault tree analysis is that it allows to analyse failure in an organized and structured graphical manner. All components, faults, conditions, as well as theirs interactions, are presented in standardized and very easy to understand and trace graphical notation. Fault tree construction is generally a complicated and time consuming task. The first step in constructing fault tree is to determine the single top event. Then the sub events are analysed and fault tree is gradually expanded. The expanding of fault tree is stopped when there is no more components needed to develop nor more sufficient data exists [19].

The authors analyse events on the B class LCPS system. The top event of fault tree is a collision caused by drivers entering the level crossing before the oncoming train.

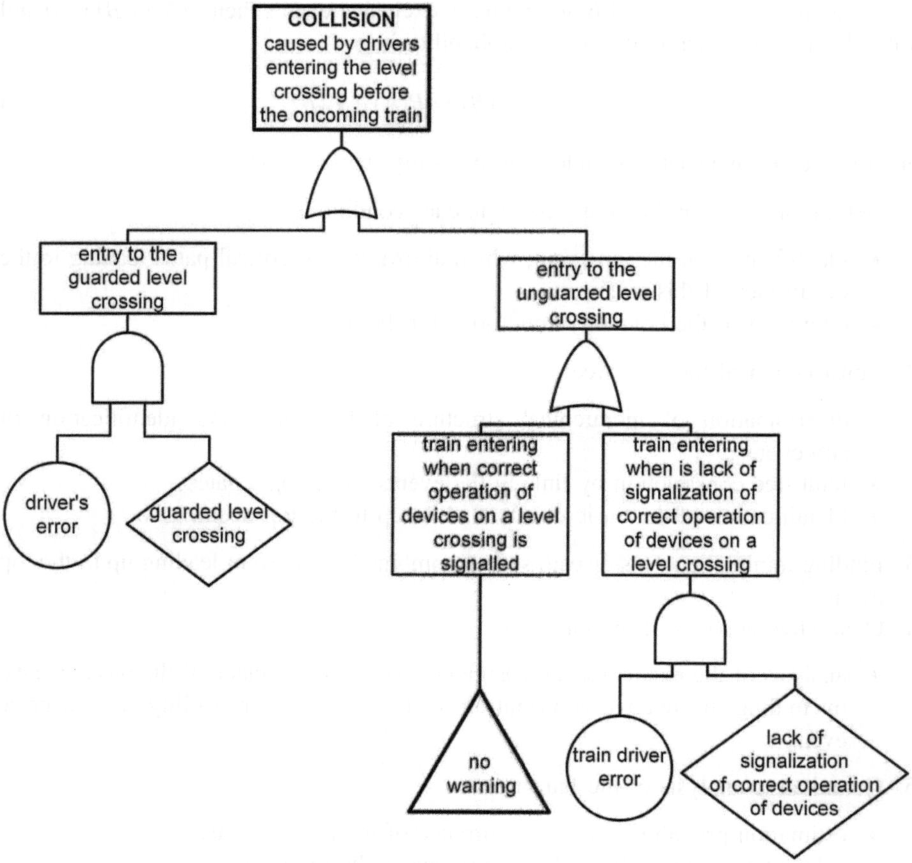

Fig. 2. Top event of LCPS fault tree

This top event is related to the two main causes: the entry into the guarded level crossing due to driver's error and entry into the unguarded level crossing as a result of the LCPS system malfunction. Figure 2 illustrates the top event of LCPS fault tree. The term "guarded level crossing" means that the railway signalling is switched on and the barriers are lowered, while "train arrival" means that the train enters the level crossing zone without speed limit.

Then "no warning" intermediate event was analysed, as is shown in Fig. 3. This may be caused due to controllers' errors in the central control unit or incorrect operation of the level crossing warning shields in combination with a group of such reasons as: the lack of detection of an approaching train, no reaction of the control system, malfunctions of the actuating devices.

The last group of the events under consideration in the fault tree is "malfunctions of the actuating devices". In the analysis was indicated that a reason for this failure could be malfunction of the turnpike and road signals damages, as shown in Fig. 4.

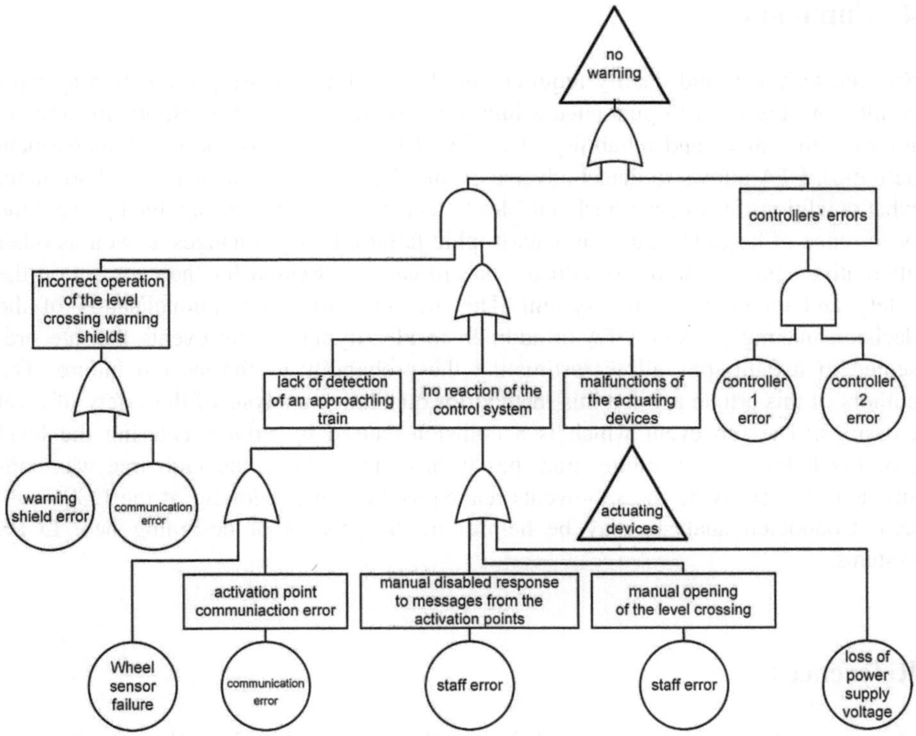

Fig. 3. Fault tree of "no warning" event

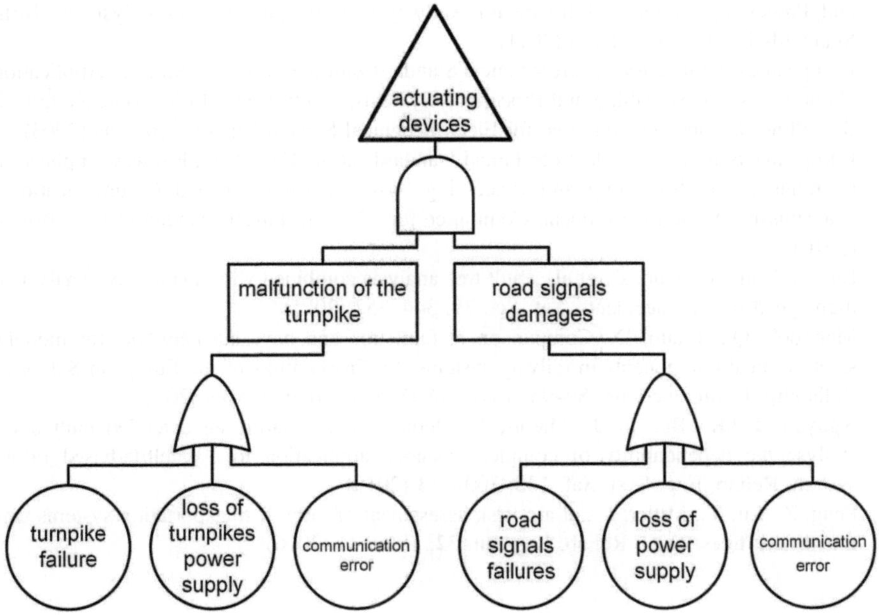

Fig. 4. Fault tree of "malfunctions of the actuating devices" event

4 Summary

Specific technical and quality requirements for the level crossing protection systems result from the need to guarantee a high level of safety in rail transport. In order to improve the safety and reliability of LCPS, it is necessary to apply risk assessment activities. FTA allows to detect adverse events. Determining whether, and if so under what conditions may occur crash, enables to take remedial measures, which prevent the occurrence of hazards leading to catastrophic failure. Design changes as well as other alternative actions leading to reduce a hazard can be assessed for their impact on the safety and reliability of the system. This in turn allows for rationalisation of the decision making process. FTA in addition to identifying system events that are presented in a fault tree, allows estimating the probability of the system failure. The authors of this article applied this method for qualitative analysis of the safety of level crossings. The top event which is a collision caused by drivers entering the level crossing before the oncoming train has been defined. Next, the fault tree was constructed by identifying the sub-events related to the malfunctioning of the LCPS system. Conducted analysis may be helpful in the process of designing new LCPS systems.

References

1. European Committee for Electrotechnical Standardization: EN 50126: Railway Applications. The Specification and Demonstration of Reliability, Availability, Maintainability and Safety. European Committee for Electrotechnical Standardization, Brussels (1999)
2. British Standards Institution: EN 50128: Railway Applications. Communication, Signalling and Processing Systems. Software for Railway Control and Protection Systems. British Standards Institution, London (2011)
3. European Committee for Electrotechnical Standardization: EN 50129: Railway Applications. Communication, Signalling and Processing Systems. Safety Related Electronic Systems for Signalling. European Committee for Electrotechnical Standardization, Brussels (2003)
4. European Committee for Electrotechnical Standardization: EN 50159: Railway Applications. Communication, Signalling and Processing Systems. Safety-Related Communication in Transmission Systems. European Committee for Electrotechnical Standardization, Brussels (2010)
5. Liu, P., Yang, L., Gao, Z., et al.: Fault tree analysis combined with quantitative analysis for high-speed railway accidents. Saf. Sci. **79**, 344–357 (2015)
6. Mahboob, Q., Straub, D.: Comparison of fault tree and bayesian networks for modeling safety critical components in railway systems. In: Proceedings of the European Safety and Reliability Conference, pp. 89–95. Taylor & Francis group, Troyes (2011)
7. Nguyen, T.P.K., Beugin, J., Marais, J.: Method for evaluating an extended fault tree to analyse the dependability of complex systems: application to a satellite-based railway system. Reliab. Eng. Syst. Saf. **133**, 300–313 (2015)
8. Peng, Z., Lu, Y., Miller, A., et al.: Risk assessment of railway transportation systems using timed fault trees. Qual. Reliab. Eng. Int. **32**, 181–194 (2016)

9. Bester, L., Toruń, A.: Modeling of reliability and safety at level crossing including in polish railway conditions. In: Mikulski, J. (ed.) Telematics - Support for Transport. CCIS, vol. 471, pp. 38–47. Springer, Heidelberg (2014)
10. Burdzik, R., Nowak, B., Młyńczak, J., Deuszkiewicz, P.: Analysis of the detection and crossing signaling system in safety terms. Diagnostyka **17**(4), 65–72 (2016)
11. Mikulski, J., Młyńczak, J.: Railroad level crossing - technical and safety trouble. In: Weintrit, A., Neuman, T. (eds.) Transport Systems and Processes, pp. 69–76. CRC Press, London (2011)
12. Kuśmińska-Fijałkowska, A., Łukasik, Z.: Information and communication technologies in the area with a complex spatial structure. In: Weintrit, A., Neuman, T. (eds.) Information, Communication and Environment, pp. 131–134. CRC Press, London (2015)
13. Lewiński, A., Perzyński, T.: The reliability and safety of railway control systems based on new information technologies. In: Mikulski, J. (ed.) Transport System Telematics. CCIS, vol. 104, pp. 427–433. Springer, Heidelberg (2010)
14. Łukasik, Z., Ciszewski, T., Młyńczak, J., Nowakowski, W., Wojciechowski, J.: Assessment of the safety of microprocessor-based semi-automatic block signalling system. In: Macioszek, E., Sierpiński, G. (eds.) Contemporary Challenges of Transport Systems and Traffic Engineering. LNNS, vol. 2, pp. 137–144. Springer, Switzerland (2017)
15. Nowakowski, W., Łukasik, Z., Bojarczak, P.: Technical safety in the process of globalization. In: 16th International Scientific Conference Proceedings Part IV: Globalization and Its Socio-Economic Consequences, pp. 1571–1578. University of Zilina, Zilina (2016)
16. International Electrotechnical Commission: International Standard IEC 61025: Fault Tree Analysis. International Electrotechnical Commission, Geneva (2016)
17. Wells, G.L.: Hazard Identification and Risk Assessment. The Institution of Chemical Engineers, Rugby (1996)
18. Rausand, M., Hoyland, A.: System Reliability Theory: Models, Statistical Methods and Applications. Wiley, New Jersey (2004)
19. Sihite, J.F.: Failure Analysis of Power Transformer Based on Fault Tree Analysis. Kyoto University, Kyoto (2013)

Determinants of the Development of Transport Systems in European Union

Transport Planning - It's Not Rocket Science

David Williams(✉)

Geography and Environmental Management Department,
University of the West of England, Bristol, UK
david23.williams@uwe.ac.uk

Abstract. When new transport infrastructure is designed and modelled we look at how it is used collectively, rather than at the individual level. However when we try change the behavior to alternative modes of transport we look at changing the individuals' behavior. This paper provides a summary of the three tools available to governments to create change to how people travel: regulation/restriction, providing new infrastructure and Voluntary Travel Behavior Change. The paper explores each option and provides an explanation of how social practice theory can provide a new way of understanding the meanings associated with how we travel that 'lock-in' travel by car. It is possible to see that a wider interpretation of transport solutions and understanding of the concept of induced demand it is possible to deliver transport solutions that change the norm towards more sustainable modes of travel, reducing the burden on the individual to change in a system that is designed for travel by car.

Keywords: Social practice theory · Behavior change · Behavioral economics · Voluntary travel behavior change

1 Introduction

I was once told that: "transport planning is not rocket science: it is far more complex than that. The rocket either goes up or it doesn't. With transport there are many variables at play at any one time to make a transport network function" [1]. I was told this by a transport modeler, who was a former rocket scientist and therefore believed what he said. We try to understand and predict how and why people travel in certain ways using a complex suite of transport models to design and test the impact of new infrastructure. In many cases models are used to design infrastructure to meet both existing and future demand of transport to ensure our towns and cities function properly. This is to enable economic growth and prosperity. All transport models look at collective movement of people and vehicles developing an origin and destination trip matrix of journeys that will pass through the modelled system. To date no transport models have been successful in capturing the complexity of multi-destination journeys, such as taking children to school, that make up much of the rush hour traffic. New tools, such as Bluetooth data [2] are starting to unlock this complexity, but the majority of models still use origin and destination matrices to plan and design changes to the transport system.

When potential new infrastructure is tested we use collective models to predict their how they will work. At the same time we send conflicting messages about behavior

E. Macioszek and G. Sierpiński (eds.), *Recent Advances in Traffic Engineering for Transport Networks and Systems*, Lecture Notes in Networks and Systems 21,
DOI 10.1007/978-3-319-64084-6_11

change to reduce demand on the transport network. These are based predominantly on psychological methods of behavior change. These 'softer' methods are designed to reduce demand on travel and are called Voluntary Travel Behavior Change (VTBC) methods. These have been trialed extensively across the developed world to varying degrees of success [3]. The largest scale delivery of VTBC initiatives took place in the United Kingdom between 2011 and 2015 [4], where the UK government invested £540 m (€640m*/US$700m* {*May 2017}) into delivering these initiatives across England [5]. The results were varied with the UK's Department for Transport indicating that a variety of measures were required to increase bus use, cycling and walking trips including: integrated ticketing, cycle training and improved walking and cycling infrastructure [5]. The primary problem with the Department for Transport's schemes was that whilst they created the opportunity for change, they did not provide a means of forcing this change to happen [6]. The second problem Secondly many of the factors that influence how people travel sit outside of the control of transport authorities, such as work and family commitments, the sunk costs of automobile ownership and expectations of how and when people should travel [6, 7]. This means that other factors play a significant role in how and why people decide to travel the way they do and this is not captured by individual, psychology based behavior change methods used to aid people in changing the way they travel.

This paper therefore explores the potential for creating change through collective behavior change based on sociological understandings of how and why people travel, rather than psychological approaches. The paper sets out an overview of how and why psychological methods have come to dominate this field, an explanation of the potential benefits that can be obtained by viewing travel through a sociological lens. The paper will conclude with a comparison between the two approaches and highlight the benefits of using a sociological approach to understanding how we design to design transport networks.

2 Behavior and Transport

Governments, both locally and nationally and their agencies often seek to change the behavior of people for many reasons, such as health and protecting the environment [8]. This means that governments play a significant role in how we act every day. The two primary means of influencing behavior are regulatory measures and non-regulatory measures [9]. The delivery of the VTBC schemes in the UK was based on the Nuffield Council for Bioethic's 2007 'Ladder of Interventions', as shown in Fig. 1 [10]. The ladder starts off with a laissez-faire approach, of providing information and enabling choice, which the UK VTBC schemes were based on [10], through to restriction and elimination of choice.

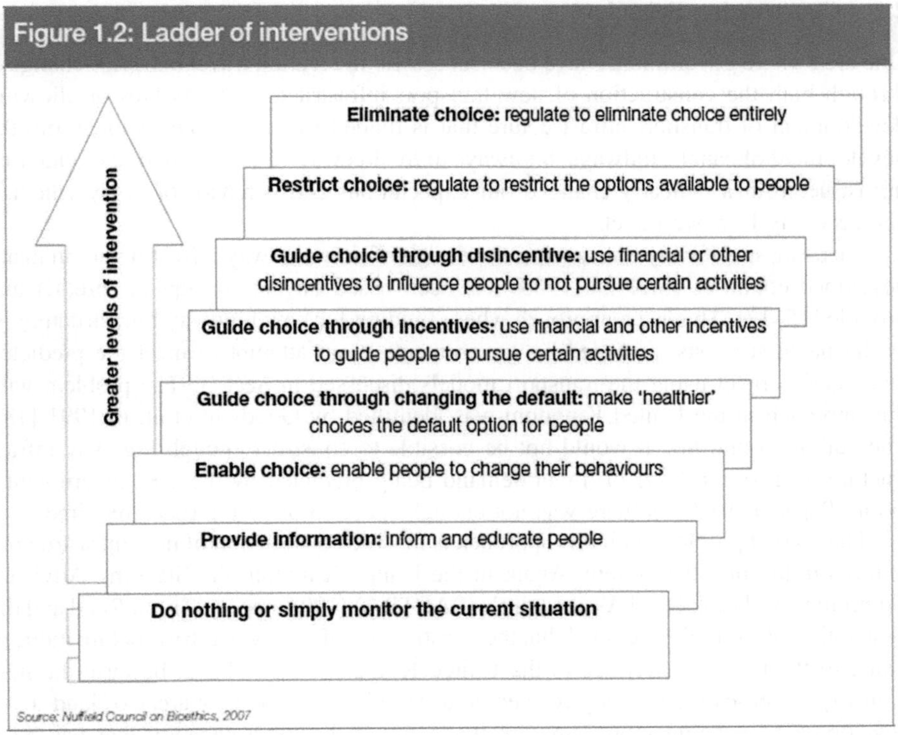

Fig. 1. Ladder of interventions (Source: Based on [11])

2.1 Restricting Behavior

Governments can use regulatory policies such as legislation and taxation to eliminate, restrict or disincentivise a choice [6]. Restrictive legislation can influence the social norms that exist around a habit, action or certain type of behaviour. Travelling by car is controlled by strict legislation and varying degrees of taxation across the world as a means of controlling who drives, how they drive, what they drive, where they drive and when they drive. An example of how people drive being influenced by the government is the practice of wearing a seatbelt. In the United Kingdom prior to this practice becoming compulsory only 40% of people wore a seatbelt when driving, compared to 93% once the law changed [12]. This is an example of a government altering behaviour for health reasons, for the benefit of people wishing to travel by car.

Using another example from the United Kingdom, the UK Government introduced a new taxation system for vehicles in 2001 designed to reduce Carbon Dioxide (CO_2) emissions. This meant that the level of CO_2 emitted by the vehicle related to the level of tax paid by the driver of the vehicle [13]. This helped to reduce CO_2 emissions from cars, as drivers opted for vehicles that were cheaper to insure [14].

2.2 Constructing Behavior

One area where governments have been successful in creating travel behavior change is through both the construction of new transport infrastructure themselves or allowing development of transport infrastructure that is funded by the private sector. From the development of canals, railways, highways to the freeways the construction of enabling infrastructure has radically changed our expectations and behavior of firstly whether, and secondly how we travel.

Focusing specifically on car trips, the design of new highway infrastructure in many developed countries since the 1960s has been based on the concept of 'predict and provide [15–17]. This is an approach where funding for new highway infrastructure is, as the name suggests, provided by governments in an attempt to meet the predicted demand for travel using the transport models discussed in Sect. 1. The problem with this approach in the United Kingdom was identified by Goodwin et al. in 1991 [18]. The authors found that it would not be possible to construct enough highway infrastructure to meet the level of travel demand being predicted by the UK government's White Paper in 1989, as there was not enough space to build the roads required.

The second problem with this approach is that the construction of new infrastructure induces traffic into the system. Again in the United Kingdom the Standing Advisory Committee on Trunk Road Assessment's (SACTRA) (1994) report 'Trunk Road and the Generation of Traffic' concluded that the construction of highway infrastructure induced traffic to the highway network in the United Kingdom [19]. This is because the new highway increases the capacity for people to travel and opens new areas of land up to development, meaning more trips are made filling the extra capacity that has been provided. SACTRA's findings were supported 23 years later by the Campaign for the Protection of Rural England's (CPRE) 2017 paper 'The Impact of Road Projects on England', which also found that road building also had little impact on congestion due to the issue of induced demand [20]. The phenomena of induced demand is not only found in the United Kingdom, but across the world due to land being opened up to development and an expectation or need to travel by car [21–23]. The construction of highway infrastructure and the associated development has shaped the way people behave and how and when they travel, and this has left many governments with an issue of how to reduce demand for the very infrastructure they have provided.

2.3 Changing Behavior

For neo-liberal governments, based on market forces it is essential that you provide choice for individuals to 'make the right decisions'. This is the bases of neo-classical economics, where everyone in the system is a 'rational actor' [24]. This has led to the development of policies that are designed to give choice and support individuals in the decision making process [25]. The belief is that outright restriction through legislation leads to people attempting to circumvent the rules. Placing restrictions, such as road user charging on an individuals' ability to travel is often met with fierce opposition [26]. This means that VTBC schemes have been trialed that are based on psychological approaches to behavior. The most successful of these is called behavioral economics.

Behavioral Economics, also known as nudge theory, has become increasingly popular with national governments who are looking to change behavior in a variety of ways. Behavioral economics moves away from the neo-liberal economic understanding of behavior and understands that people do not always act rationally [27]. The approach is designed to change behavior through information, incentives and contexts [27]. These simple, relatively small-scale changes then change individuals' behavior towards the type of behavior desired by the policy maker.

The primary problem with the use of behavioral economics is that what is required to change travel behavior away from travelling by private car is vastly different from 'nudging' behavior towards making people reuse their towels in a hotel, an example cited by the authors of 'Nudge: Improving decisions about health, wealth and happiness' [28]. The choice to own and run a car is dictated by a multitude of factors including: sunk costs, societal expectations, family commitments, work pressures and the lack of alternative options due to land-use planning based around the car [6]. Providing information, sustainable transport infrastructure through VTBC and incentives to change are unlikely to be successful, as many people have based their lifestyle around an expectation of being able to travel by car. The element of choice and lack of restriction means that for many people travelling by car will remain the most logical choice. This means that we need to look at transport and land-use planning in a different, more collective way. This collective approach is not unusual for creating new infrastructure, based on models of mass behavior, so why should it be any different for attempts to reduce travel by car?

3 Sociological Approaches to Changing Travel Activity

The primary problem is that behavioral approaches to creating change focus on the individual and expect them to create a change in a system that has not been designed to make that choice the default. This makes creating that change difficult if you focus on the individual. It is therefore worth exploring alternative approaches to understanding how and why we do things in a certain way. Social practice theory is on such alternative approach that focuses on the action being performed, rather than the practitioner performing it. This allows for the wider perspectives of travel to be incorporated into the decision making exercise. The individual is therefore no longer the unit of inquiry [29, 30].

The problem with social practice theory is explain in [31]: "social theories do not lead directly to prescriptions for actions". Due to the relative complexity of practices it can often be difficult for policy makers to come up with a solution that will enable change and this is why social practice theory has not been used to deliver behavior change initiatives within the transport sector [31]. It is however possible to identify tools that could be useful for delivering infrastructure in a new way.

3.1 Three Elements of a Social Practice

The scientific work [31] provided a simple model to explain what a practice is, as shown in Fig. 2. A practice can be broken down three core elements:

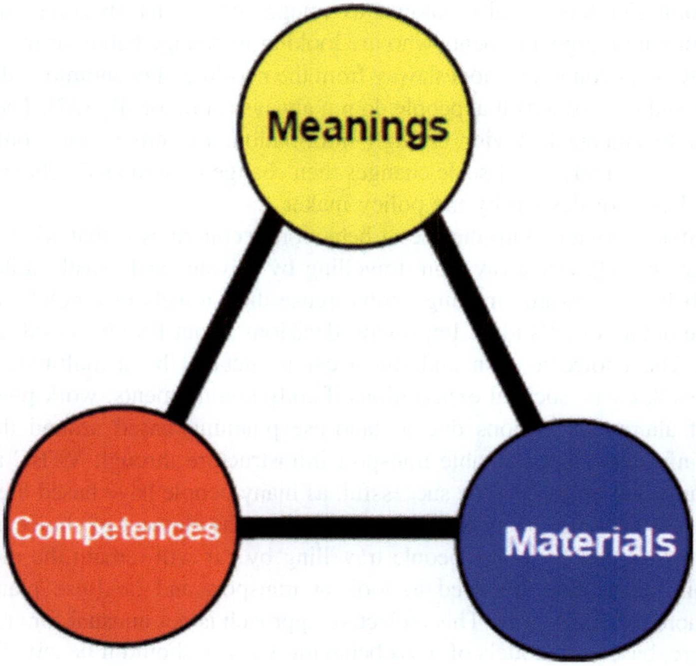

Fig. 2. The 3-Elements Model (Source: Based on: [6, 31])

- materials: things, technologies, tangible physical entities and the things of which objects are made,
- competences: skill, know-how and technique,
- meanings: symbolic, ideas and aspirations [31].

By focusing on what practice is being performed, rather than who is performing it we can begin to understand why people so many people travel by car. Focusing on the materials for travel it is clear that the construction of new transport infrastructure enables the practice of travelling to be performed. Highways have existed for thousands of years, but it is only within the past 130 years that these have started to be designed for use by motorized transport. This is because of the rise in availability of private motor vehicles, another material that has become available to the majority of the population. The construction of this infrastructure has been incremental and has allowed for a change in the number of activities that can be performed by a practitioner at a variety of locations within a specific day [19].

The second element that is required for a practice to be performed is competence to undertake it. This is known as practical understanding [32] and this is learnt through either formal training, such as learning to drive, or informal training through everyday life. The more we practice something the better we get at it.

The third element within the model is meaning. Meanings within practices can vary in different locations. For example drinking tea is performed differently in the United Kingdom and Japan [29], with local customs and meaning not always governed by formal rules [33]. For the practice of driving this has become formally imbedded within western society through representation in the media, friends and family and the existing infrastructure that is provided for travel [6]. It is therefore important that we understand the meanings that become associated with the materials and competences that we provide for travel.

3.2 Changing Practices

Creating change to practices can evolve over time, or change quickly due to the introduction of changes to the materials available, the skills to use them and the meanings associated. When the links between these are broken then a new practice is formed. This is what happened with the evolution of the motor car, as a new material was available for travel, people who could afford to, switched from other modes, or travelled more than they had previously. This led to the design of highways based around car travel to the design of homes and buildings to accommodate motor vehicles [34]. This then locks in the practice so that it becomes normal [35]. When viewing how people travel through the 3-Elements model it is possible to see the difficulty of trying to change the behavior of an individual [or practitioner], when the practice has become normalized and therefore difficult to break.

Focusing on VTBC initiatives as a means of changing the way people travel, it is clear that a variety of methods have been successful in certain areas of the UK following their delivery between 2011 and 2015 [5], but it is uncertain how long the benefits will last. Bonsall [3] found that internationally, evidence of the long-term success of VTBC schemes to create lasting change was difficult to find [8]. This is because many of the VTBC schemes did not include evaluation programmes. In addition it is difficult to demonstrate the long-term benefits of some VTBC initiatives such as cycle training for children within an evaluation time frame, as the benefits may take many years to be realized [6]. This highlights the difficulties of trying to create change. In addition the practice of travelling sits within a whole web of other practices based around how we live our lives. Changing travel habits would require a shift in the meanings around child-care, shopping and work commitments that have built up around being mobile and visiting multiple destinations within a day.

All VTBC schemes are very good at providing two of the three elements of travel: materials and competences. These are provided through the provision of public transport, new cycle and walking infrastructure and training for children and adults to use it effectively. What is lacking is any attempt to change the meaning of how to travel, which often has to be made through restriction of an existing behavior. The meanings of VTBC delivery in the UK did not change with the additional funding for transport, especially as the UK Government provided nine-times as much funding for highway infrastructure for cars during the same period as the LSTF [36]. This provision of highway infrastructure continues to influence the meanings associated with travelling around the use of the car, rather than sustainable transport.

Restriction is one of the key tools available to transport practitioners if they want to change the way people travel. The removal of highway space available for cars has been seen across the world to make traffic 'disappear' from the wider transport network [37].

This is because the meanings associated with this space have changed, meaning the network is only available to public transport, cyclists or pedestrians rather than the free-flow of traffic. What was also interesting was that those trips did not re-appear elsewhere in the network in any of the 80 case studies reviewed [37]. This removal of a material had altered the practices of travel, with people either deciding to travel by a different mode or not to travel at all. It is clear from this example that if we think about the message, or meaning that we are trying to send from the transport network we provide it is possible to change how people travel without relying on individual approaches to behavior change within a system that makes this change difficult.

4 Conclusion

The use of behavioral economic approaches to behavior change are a useful, but limited tool available when creating change. This is because they require an individual to break the norms of society to travel in a different way. The delivery of VTBC schemes provide some benefit to creating a change to how people travel, but at present there is still not definitive proof that they are successful at permanently altering behavior in the long-term. This is because many factors, or other practices in daily life influence how we decide to travel.

Transport planners are used to exploring change through complex models that explore the costs and benefits of providing new transport infrastructure. This is often focused on providing new highway infrastructure and how this will influence the amount of traffic on the network. There is a considerable wealth of data that proves that providing highway infrastructure for car trips has no economic benefit, induces extra traffic to the system and rarely, if ever, solves congestion. This collective approach to viewing travel is not rare to transport planners, so there is a need to use this when trying to change behavior.

Shove et al.'s 3-elements model, provides a way of viewing any transport intervention in a new way that incorporates an understanding of the meanings you create through the solution you provide. Build more roads and you get more traffic. Close/restrict roads and provide alternative means of travel and you get less traffic. It is not rocket science.

References

1. Fox, K.: Email to David Williams. Via Linkedin (2014)
2. Crawford, F.: A Clustering Method Based on Repeated Trip Behaviour to Identify Road User Classes Using Bluetooth Data. Institute for Transport Studies, Leeds (2017)
3. Bonsall, P.: Do we know whether personal travel planning really works? Transp. Policy **16**, 306–314 (2009)
4. Department for Transport. Government Announces Plans for a New Transport Fund. http://acttravelwise.org/news/1597/
5. Department for Transport. Local Sustainable Transport Fund 'What Works' Report. https://www.gov.uk/government/publications/local-sustainable-transport-fund-what-works

6. Williams, D.: Social Practice Theory and Sustainable Mobility: An Analysis of English Local Transport Planning as a System of Provision. University of the West of England, Bristol (2014)
7. Sheller, M., Urry, J.: The city and the car. Int. J. Urban Reg. Res. **24**, 737–757 (2000)
8. Bonsall, P., Conner, M., Darnton, A., Marsden, G. Thinkpiece, I.: Influencing Individual Citizens. Final Report from the Institute for Transport Studies. London Department for Transport, London (2009)
9. Behaviour Change: Science and Technology Select Committee 2nd Report of Session 2010–2012. http://www.publications.parliament.uk/pa/1d201012/ldselect/ldsctech/179/179.pdf
10. Department for Transport. Creating Growth, Cutting Carbon: Making Sustainable Local Transport Happen. https://www.gov.uk/government/publications/creatinggrowth-cutting-carbon-making-sustainable-local-transport-happen
11. Nuffield Council on Bioethics, Contains Public Sector Information Licensed Under the Open Government Licence v3.0. http://www.nationalarchives.gov.uk/doc/open-government-licence/version/3/
12. Department for Transport. Seat Belt Wearing Rates. http://www.gov.uk/government/uploads/system/uploads/attachment_data/file/8900/seatbelt-rates.pdf
13. Potter, S., Atchulo, A.: A Review of Ten Years of CO_2—Based Company Car Taxation: Impact and Potential. Oxford University, Oxford (2013)
14. Department for Transport. Vehicle Licencing Statistics. https://www.gov.uk/goverment/uploads/system/uploads/attachment_data/file/608374/vehicle-licensing-statistics-2016.pdf
15. Dudley, G., Richardson, J.: Why Does Policy Change? Lessons from British. Transport Policy 1945–99. Routledge, London (2004)
16. Vigar, G.: The Politics of Mobility: Transport, the Environment and Public Policy. Taylor & Francis, London (2002)
17. Goodwin, P.: 'Peak Car'—Where Did the Idea Come From? And Where Is It Going?. University of the West of England, Bristol (2012)
18. Goodwin, P., Hallett, S., Kenny, F., Stokes, G.: Transport: The New Realism. University of Oxford, Oxford (2012)
19. The Standing Advisory Committee on Trunk Road Assessment: Trunk Roads and the Generation of Traffic. Crown, London (1994)
20. National Archives Website. http://webarchive.nationalarchives.gov.uk/
21. Sloman, L., Hopkinson, L., Taylor, I.: The Impact of Road Projects in England. Campaign to Protect Rural England, London (2017)
22. Handy, S.: Smart growth and the transportation land use connection: what does the research tell us? Int. Reg. Sci. Rev. **28**, 146–167 (2005)
23. Gorham, R.: Demystifying Induced Travel Demand. Sustainable Urban Transport Technical Document. Federal Ministry for Economic Cooperation and Development, Eschborn (2009)
24. Litman, T.: Generated traffic and induced travel: implications for transport planning. ITE J. **71**, 38–47 (2017)
25. Earle, J., Cahal Moran, C., Ward-Perkins, Z.: The Econocracy: The Perils of Leaving Economics to the Experts. Manchester University Press, Manchester (2016)
26. Disney, K., Le Grand, J., Atkinson, G.: From Irresponsible Knaves to Responsible Knights for Just 5p: Behavioural Public Policy and the Environment. Behavioural Public Policy. Cambridge University Press, Cambridge (2013)
27. Shaw, J., Docherty, I.: New Deal or no new deal? A decade of 'sustainable' transport in the UK. In: Docherty, I., Shaw, J. (eds.) Traffic Jam: Ten Years of 'Sustainable' Transport in the UK, pp. 3–26. Policy Press, Bristol (2008)

28. Dolan, P.: Influencing the financial behaviour of individuals: the mindspace way. In: Oliver, A. (ed.) Behavioural Public Policy, pp. 191–208. Cambridge University Press, Cambridge (2013)
29. Thaler, R., Sunstein, C.: Nudge: Improving Decisions about Health, Wealth and Happiness. Penguin, London (2008)
30. Chatterton, T.: An Introduction to Thinking about 'Energy Behaviour': A Multi Model Approach. University of the West of England, Bristol (2011)
31. Shove, E., Pantzar, M., Watson, M.: The Dynamics of Social Practice: Everyday Life and How it Changes. Sage, London (2012)
32. Cairns, S., Harmer, C., Hopkin, J., Skippon, S.: Social perspectives on travel and mobilities: a review. Transp. Res. Part A **63**, 107–117 (2014)
33. Reckwitz, A.: Toward a theory of social practices: a development in culturalist theorizing. Eur. J. Soc. Theory **5**, 243–263 (2002)
34. Schatzki, T.: Social Practices: A Wittgensteinian Approach to Human Activity and the Social. Cambridge University Press, Cambridge (1996)
35. Morrison, K., Minnis, J.: Carscapes: The Motor Car, Architecture, and Landscape in England. Yale University Press, London (2012)
36. Shove, E.: Comfort, Cleanliness and Convenience: The Social Organization of Normality. Berg Publishers, Oxford & New York (2003)
37. Williams, D., Chatterton, T., Parkhurst, G.: Assessing the Local Sustainable Transport Fund Submissions in Relation to Disruption. In: Proceedings of the 45th Universities' Transport Study Group. Oxford University, Oxford (2013)

Parameters of Bus Lines Influencing the Allocation of Electric Buses to the Transport Tasks

Krzysztof Krawiec(✉) and Marcin Jacek Kłos

Faculty of Transport, Silesian University of Technology, Katowice, Poland
{krzysztof.krawiec,marcin.j.klos}@polsl.pl

Abstract. According to the transport policy of the European Union electric buses are being introduced on a broader scale. In order to induct electric buses in service, a public transport company has to decide on the allocation of electric buses to specific bus lines. This decision should not be taken in subjective way but it should rather be supported by the scientific methodology that is to determine the parameters of bus lines, which may affect the allocation of buses to the transport tasks. The result of the article is a selection of the most important parameters of bus lines that influence the allocation of electric buses to the transport tasks.

Keywords: Battery electric bus · Public transport · Transport modelling

1 Introduction

In line with the transport policy of the European Union there is an increase of interest in electric vehicles in public transport. Electric buses have many advantages such as no local environmental emissions and better energy efficiency than conventional Diesel buses. This causes the lines serving the city centers which are characterized by frequent stops and low speeds are perfect to be operated by electric buses [1–3]. However, the main problem with electric buses is that their range is limited - depending on local conditions - from 150 up to 300 km. Hence, there is a need to re-charge electric buses during the time of its operation. On the one hand this problem can be solved by improving battery technologies or on the other hand by optimization of locations of battery re-charging stations [4–7].

The process of the implementation of electric buses to public transport company is complex from an operational and economic point of view [8]. In order to induct electric buses in service, a public transport company has to decide on the allocation of electric buses to specific bus lines. There is no doubt that the parameters of public transport lines have a large impact on the performance of electric buses. Depending on the route characteristics (urban, sub-urban, inter-urban, etc.) the disadvantages of electric buses can be either more or less perceptible from the public transport operator point of view. It is therefore necessary to select the bus lines that are most suitable to be operated by electric buses.

E. Macioszek and G. Sierpiński (eds.), *Recent Advances in Traffic Engineering for Transport Networks and Systems*, Lecture Notes in Networks and Systems 21,
DOI 10.1007/978-3-319-64084-6_12

2 Methods of Selection of Variables in Transport Studies

Transport systems are characterized by a wide range of relationships and links between features and factors affecting its functioning. This is confirmed by arguments presented i.e. in publications [9, 10] presenting a holistic description of the functioning of transport systems. The systems approach translates to a large extent on the formation of problems associated with taking into account all the features and factors in the models which are to be constructed to describe the reality as precisely as possible.

Accurate representation of modelled reality binds very often with the necessity to perform a large number of empirical studies. In many cases it is hardly possible due to financial, technical and organizational limitations for the research teams - especially analyzing phenomena and processes occurring in the transport systems. There is thus often a need to limit the number of variables so as to influence the deterioration of a models' quality in a minimal extent.

In general, the methodology of limiting the variables describing the functioning of transport systems can be performed in the two following stages:

- stage A - elimination of variables due to the specific conditions of studied phenomena and processes. In this step a subjective assessment of the suitability of the studied traits in the description of the model is conducted. An author of study eliminates the variables through an analysis without the use of mathematical (statistical) tools,
- stage B - elimination of these variables, which have not been removed in the stage 1. These variables have been subjected to empirical research and their eliminations is done with the use of mathematical (statistical) tools.

Application of above-named methodology is presented in [11, 12] that deal with the issue of capacity of various road junctions types. So, the aim of this article is a description of stage A which consists of four steps:

- step 1. Selection related to the assumptions of the system of public transport operated by electric buses,
- step 2. Selection depending on the strong ties to the other defined factors,
- step 3. Selection due to the fact that this group of factors is not affected by the type of applied propulsion,
- step 4. Selection for the enhancement of the universality of the method.

This method will be applied in order to determine the parameters of bus lines which may affect the allocation of buses to transport tasks.

3 Electric Buses in Urban Public Transport

3.1 General Assumptions of the System of Public Transport Operated by Electric Buses

Electric buses are in operation for a few years and no standard rules of electric buses operation have yet been developed. The assumptions for which the elimination of the selected factors will be conducted are presented below:

Table 1. Parameters of the public transport lines

Group of the parameters	Parameter names
Parameters related to the timetable	The frequency of the line
	The regularity of the line
	The frequency of terminus stops
	The set of time spent in the terminus stops
	The number of services
	Passenger flow volume
	Operating speed
	Schedule speed
	The set of time intervals in which charging is possible
	Vehicle kilometers travelled
	Vehicle hours travelled
The parameters related to the line route	Traffic conditions
	The number of scheduled bus stops due to the timetable
	The number of stops of the bus not covered by the timetable
	Technical classes of the roads within a line route
	Locations of the line route to the city center
	The existence of dedicated bus lanes within a line route
	The existence of priorities for city buses within a line route
	Vertical profile
Operational parameters	Number of buses in operation per line
	Bus capacity
	Bus length
	Reliability of operation
	Punctuality
Technical parameters	Access to the power grid at the terminus stops
	Access to the power grid at the intermediate stops
	Reliability of vehicles
	The demand for energy (fuel consumption by Diesel buses)
Economic parameters	Costs dependent on vehicle kilometers travelled
	Costs dependent on vehicle hours travelled
	Costs dependent on the number of buses in operation

- operation of 100% electric fleet on each bus line,
- charging is possible only in the depot and in the terminus stops,
- a timetable takes into account traffic conditions,
- a timetable will not be changed after the implementation of electric buses,
- the allocation of buses to the transport tasks is optimal.

As it is clear from above, no mixed fleet operation is possible. This means that either pure electric fleet is considered to be introduced on the analyzed line or no changes are to be done. Among the many charging technologies such as plug-in charging, pantograph and inductive charging [13–15] but also battery exchange systems [16] and fuel-cell-based hydrogen energy extenders [17], it is assumed that only these technologies that enable opportunity charging while the bus is stopped are taken into account. Hence, the authors do not consider en route charging technologies such as inductive charging. It is assumed that a research considering the number of passengers in buses has been conducted and the allocation of buses to transport tasks is optimal.

3.2 Identification of Parameters

On the basis of the analysis of the bibliography concerning public road transport efficiency [18], transport economics and organization of public transport [19] as well as the issue of quality and customer service in public transport [20–22], a number of parameters describing economic, operational, technical characteristics of bus lines and the parameters related to the timetable and to the line route are presented in the Table 1.

Bus lines are described by a number of parameters - to simplify the model it is thus necessary to take into account only a few of them. The above-mentioned parameters will be eliminated for its compliance with the assumptions, mutual relationships, impact of the type of drive on the parameter and the versatility of the method.

4 Determination of the Parameters of Bus Lines Which May Affect the Allocation of Buses to the Transport Tasks

4.1 Selection Related to the Assumptions of the System of Public Transport Operated by Electric Buses

In the first step, the parameters are being eliminated according to the above-mentioned assumptions. The elimination procedure is shown in the Fig. 1.

The main reason for the elimination of operating and schedule speed, the number of services as well as vehicle kilometers travelled and vehicle hours travelled, is the assumption that a timetable will not be changed after the electric buses implementation. Modifications to the timetable are not desirable by passengers therefore public transport organizers try to minimize them.

Operating speed is also eliminated due to the assumption that a timetable takes into account traffic conditions. One should notice that the traffic conditions influence the operational speed, however, on the basis of multiannual practice in organizing public

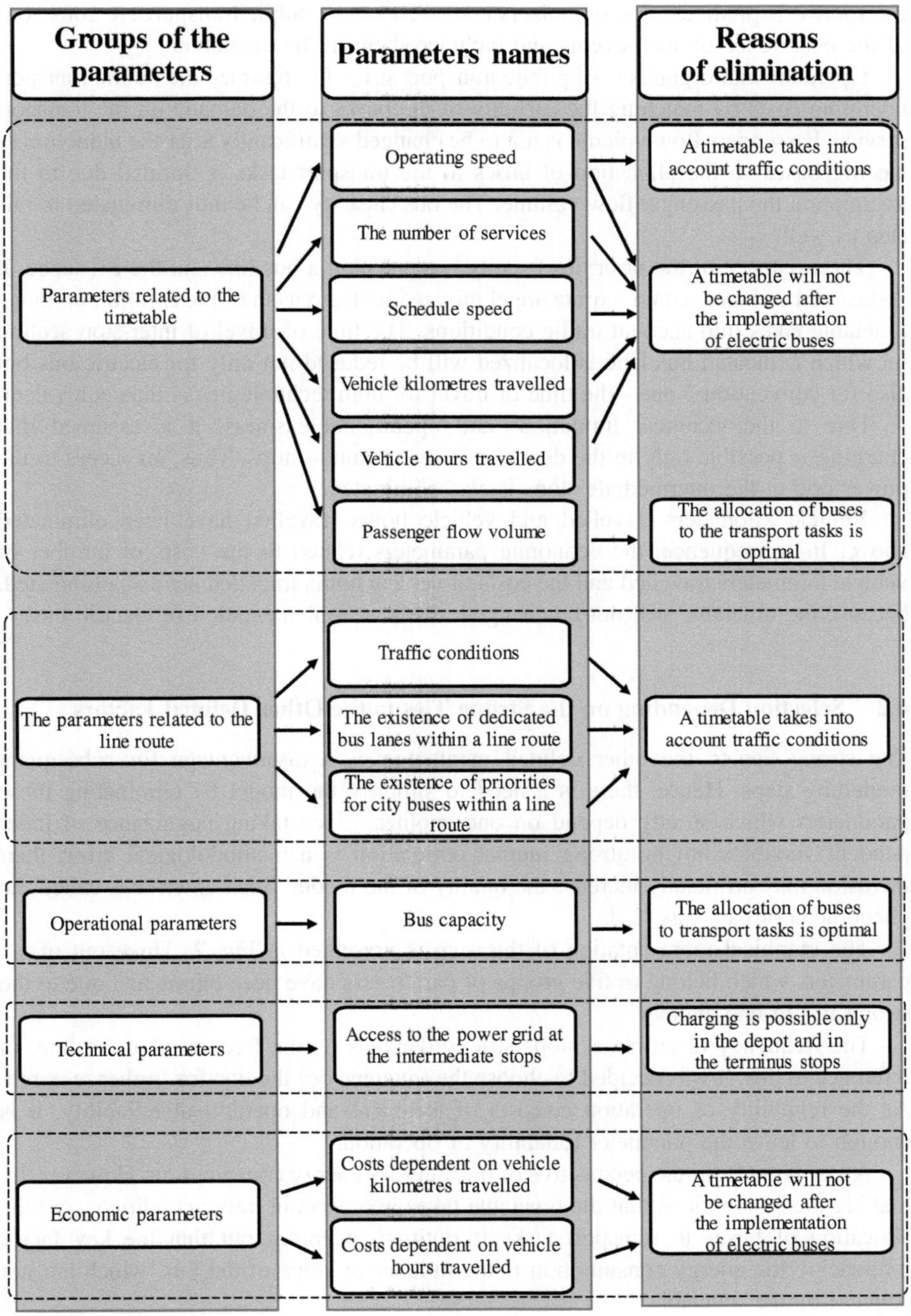

Fig. 1. Selection related to the assumptions of the system of public transport operated by electric buses

transport it is predicted that organizers and operators of public transport are conscious of the occurrence of such events and included them in the timetable.

Operators and organizers of public transport strive to minimize the public transport operating costs by matching the capacity of the buses to the demand on the transport service. Passengers flow volume is not to be changed significantly after the induction of electric buses so the allocation of buses to the transport tasks is optimal due to the assumption the passenger flow volume. The bus capacity can be thus eliminated in this step as well.

The existence of the priorities for city buses within a bus line and the existence of dedicated bus lanes within a route are eliminated by the reason of the assumption that a timetable takes into account traffic conditions. The time of travel of inter-stop section on which dedicated bus lane is localized will be reduced not only for electric bus but also for conventional one - the time of travel for both technologies is thus equivalent.

Due to the technical limitations and operational easiness, it is assumed that charging is possible only in the depot or in the terminus stops. Thus, an access to the power grid at the intermediate stops is also eliminated.

Vehicle kilometers travelled and vehicle hours travelled have been eliminated above. In consequence, the economic parameters related to the costs of number of vehicle kilometers travelled and the costs of vehicle hours travelled are also eliminated, because the timetable will not be changed after the implementation of electric buses.

4.2 Selection Depending on the Strong Ties to the Other Defined Factors

The strong ties to the other defined parameters is a disadvantage for subsequent modelling steps. Hence, there is a need to simplify the model by eliminating these parameters which strictly depend on one another. Since taking cognizance of independent variables having strong mutual correlation is a methodological error, their inclusion may artificially increase the quality of the model. Such an error is called the phenomena of catalysis [23].

The graphical representation of this step is presented in Fig. 2. Three out of six parameters which belong to five groups of parameters have been eliminated due to the strong ties to one another.

The frequency of terminus stops strictly depends on the frequency of the line. In reference to this, it was decided to choose the frequency of the line for further research. As the reliability of operation consists of technical and operational reliability, it is enough to leave the parameter reliability of operation.

A key issue from the perspective of bus range is energy consumption. However, by making the assumptions that the timetable takes into account traffic conditions and the allocation of buses to transport tasks is optimal, it comes out that the key factor influencing the energy consumption is the number of stops of the bus, which are not covered by the timetable.

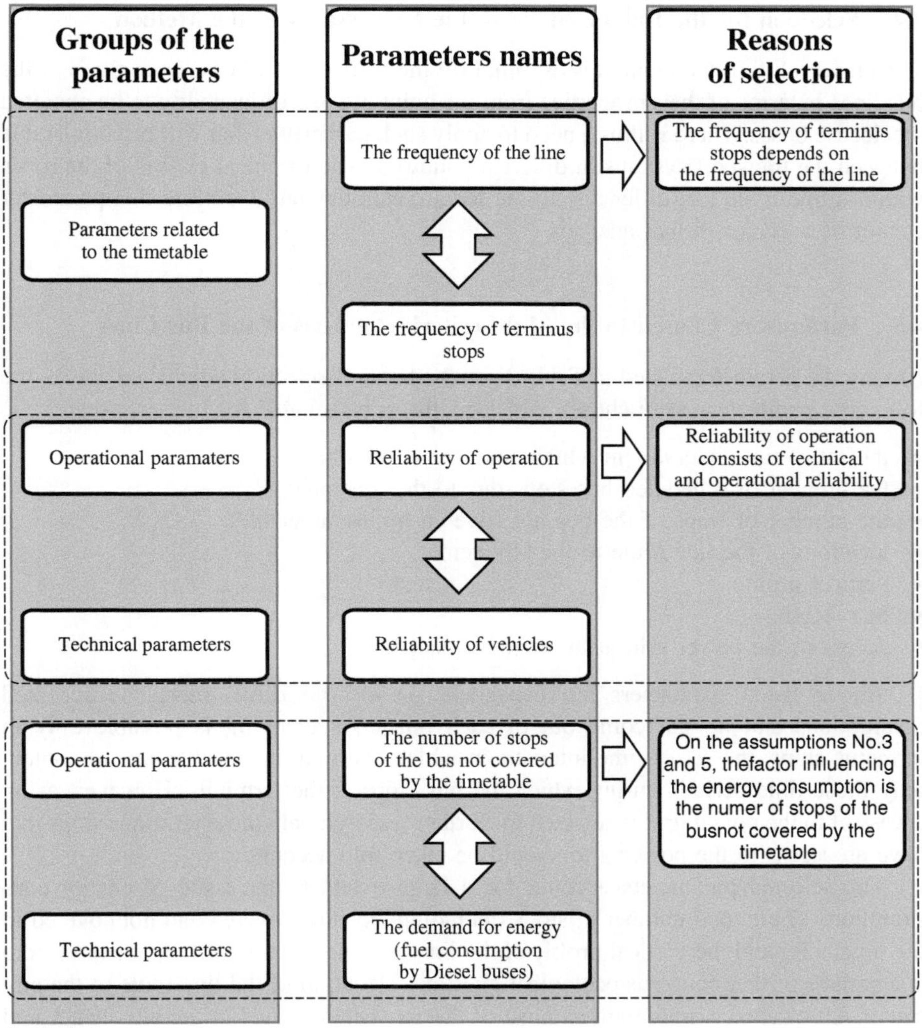

Fig. 2. Selection because of the strong ties to other defined factors

4.3 Selection Due to the Fact that This Group of Factors Is not Affected by the Type of Applied Propulsion

There are parameters of the public transport lines that will not change after the induction of electric buses. Among them, the number of buses in operation per line should be mentioned as well as the frequency of terminus stops as the timetable will not be changed after the induction of electric buses. Hence, the costs dependent on the number of buses in operation, will not be changed as well. Another parameter to be eliminated in this step is the reliability of operation. This parameter is not affected by the type of applied drive.

4.4 Selection for the Enhancement of the Universality of the Method

The universality of the method is essential for the planned multi-criteria analysis of the bus lines in terms of their properties from the point of view of its ability to be operated by electric buses. There is thus a need to apply such parameters that will be comparable between the transport systems in different countries. The technical classes of the roads within a line route can differ due to the local conditions and therefore this parameter cannot be used for further analysis.

4.5 Parameters Chosen to the Multi-criteria Analysis of the Bus Lines

Among the parameters listed in Table 1, on the basis of selection conducted above, the following parameters were chosen which seems to be reliable for further research:

- the set of time intervals in which charging is possible,
- the number of scheduled bus stops due to the timetable,
- the number of stops of the bus not covered by the timetable,
- locations of the line route to the city center,
- vertical profile,
- bus length,
- access to the power grid at the terminus stops.

Above listed parameters are dependent on electric drive, meet the accepted assumptions and do not come out of each other. As charging is possible only in terminus stops, the set of time intervals in which charging is possible should contain only those time periods during which the bus stops at the terminus. However, as an access of to the power grid is selected for further research only these terminus stops that have an access to the power grid should be taken into account.

The selected parameters account for a large extent to the issue of energy consumptions as the total number of stops (including the calls covered and not covered in the timetable) and the vertical profile of the line. Among the selected parameters, there is a parameter that concerns ecological question—location of the line route to the city canter. This factor is important as most of the passengers travel in the city center and the ecological and health benefits of reducing the impact of public transport on the environment will be the most significant.

5 Conclusions

Electric buses are more dependent on the parameters of public transport lines due to their insufficient range—it is therefore important to select the correct parameters of the bus lines influencing the allocation of electric buses to the transport tasks. Accordingly to the procedure conducted above, the parameters that have the greatest influence on the decision concerning the bus allocation to transport tasks. These parameters will provide the basis for a multi-criteria analysis of these factors to create a ranking of these bus lines which parameters are the most conductive to the implementation of electric buses.

Hence, further research will focus on the measurements of defined variables and their statistical analysis.

References

1. Lajunen, A.: Energy consumption and cost-benefit analysis of hybrid and electric city buses. Transp. Res. Part C **38**, 1–15 (2014)
2. Lajunen, A., Lipman, T.: Lifecyle cost assessment and carbon dioxide emissions of diesel, natural gas, hybrid electric, fuel cell hybrid and electric transit buses. Energy **106**, 329–342 (2016)
3. Jerram, L., Gartner, J.: Executive Summary Electric Drive Buses. Research Report. Pike Research, Colorado (2012)
4. Mohamed, M., Farag, H., El-Taweel, N., Ferguson, M.: Simulation of electric buses on a full transit network: operational feasibility and grid impact analysis. Electr. Power Syst. Res. **142**, 163–175 (2017)
5. Li, J.Q.: Transit bus scheduling with limited energy. Transp. Sci. **48**(4), 521–539 (2013)
6. Wang, Y., Huang, Y., Xu, J., Barclay, N.: Optimal recharging scheduling for urban electric buses: a case study in Davis. Transp. Res. Part E **100**, 115–132 (2017)
7. Xylia, M., Ledyc, S., Patrizio, P., Kraxner, F., Silveira, S.: Locating charging infrastructure for electric buses in Stockholm. Transp. Res. Part C **78**, 183–200 (2017)
8. Krawiec, S., Łazarz, B., Markusik, S., Karoń, G., Sierpiński, G., Krawiec, K.: Urban public transport with the use of electric buses - development tendencies. Transp. Probl. **11**(4), 127–139 (2016)
9. Jacyna, M.: Wybrane Zagadnienia Mdelowania Systemów Transportowych. Oficyna Wydawnicza Politechniki Warszawskiej, Warszawa (2009)
10. Ortúzar, J., Willumsen, L.: Transport Modelling. Willey, New York (2011)
11. Sobota, A.: Analiza Funkcjonalna Czterowlotowych Skrzyżowań Położonych na Wielopasowych Ciągach Ulic w Warunkach Nasycenia Ruchem. Uniwersytet Technologiczno-Przyrodniczy w Bydgoszczy, Bydgoszcz (2012)
12. Szczuraszek, T.: Badanie Zagrożeń w Ruchu Drogowym. Polska Akademia Nauk, Warsaw (2005)
13. Gao, Z., Lin, Z., LaClair, T.J., Liu, C., Li, J.M., Birky, A., Ward, J.: Battery capacity and recharging needs for electric buses in city transit service. Energy **122**, 588–600 (2017)
14. Kühne, R.: Electric buses - an energy efficient urban transportation means. Energy **11**(4), 4510–4513 (2010)
15. Markusik, S., Krawiec, S., Łazarz, B., Karoń, G., Janecki, R., Sierpiński, G., Krawiec, K.: The Technical and operational aspects of the introduction of electric-powered buses to the public transportation system. Logist. Transp. **3**, 41–50 (2015)
16. Kim, J., Song, I., Choi, W.: An electric bus with a battery exchange system. Energies **8**, 6806–6819 (2015)
17. Fernandez, R.A., Cilleruelo, F.B., Martinez, I.V.: A new approach to battery powered electric vehicles: a hydrogen fuel-cell-based range extender system. Int. J. Hydrogen Energy **41**(8), 4808–4819 (2016)
18. Jarboui, S., Forget, P., Boujelbene, Y.: Public road transport efficiency: a literature review via the classification scheme. Public Transp. **4**(2), 101–128 (2012)
19. Wyszomirski, O.: Transport Miejski. Ekonomika i Organizacja. Wydawnictwo Uniwersytetu Gdańskiego, Gdańsk (2010)

20. Rudnicki, A.: Jakość Komunikacji Miejskiej. Monograph 5, z. 71. Zeszyty Naukowo-Techniczne Oddziału SITK w Krakowie, Kraków (1999)
21. Starowicz, W.: Jakość Przewozów w Miejskim Transporcie Zbiorowym. Wydawnictwo Politechniki Krakowskiej, Kraków (2007)
22. Amponsah, C.T., Adams, S.: Service quality and customer satisfaction in public transport operations. Int. J. Serv. Oper. Manage. **25**(4), 531–549 (2016)
23. Gruszczyński, M., Podgórna, M.: Ekonometria. Szkoła Główna Handlowa w Warszawie, Warsaw (2004)

Possibilities for Enhancement of the Quality of Krakow's Tram Transport with Tunnel Sections

Marek Bauer(✉)

Institute of Road and Railway Engineering,
Cracow University of Technology, Cracow, Poland
mbauer@pk.edu.pl

Abstract. Presented paper compares transport capacities, travel times and speeds attained by underground trains and trams. The components of the journey time for public transport trips have been described. Differences in the travel time resulting from different lengths of track sections between stops are discussed. The physical and regression models for estimation of running time for trams were presented and compared. And then, new locations for tunnel tram routes in Krakow have been proposed. For tunnel sections, potential benefits for public transport passengers, which can be used in the planning analyzes of Krakow's transport system development, were identified.

Keywords: Transport planning · Tram · Underground · Tunnel · Travel time

1 Introduction

Rail transport constitutes the basis for transport systems in numerous cities. This stems from its high transporting capacity and on the other hand - from the high speed achieved by commonly used separation of tracks from road lanes used by other traffic users [1]. Since cities continue to grow - the discussions are still taking place on the possibility for the development of rail transport systems. The main question is: which kind of rail transport is the most efficient for the city? In many cities, take place the discussions as to, whether it is legitimate to build a new or extend existing system of public transport. Also in Krakow, what is resulting directly in the elaboration of the present paper.

Decisions should always base on the analysis of costs and benefits before the potential introduction or expansion of a system, with using a four-stages procedure (described, among others, by Hensher [2]), commonly used in the planning of transportation systems for a determination of potential effects of differentiated variants of transport investments. It is impossible to do it properly without reliable estimation of transport capacities and travel times of different means of public transport, because especially travel time has a crucial influence on modal split [3] and even slight differences may have a considerable influence on increased or decreased passenger volumes.

E. Macioszek and G. Sierpiński (eds.), *Recent Advances in Traffic Engineering for Transport Networks and Systems*, Lecture Notes in Networks and Systems 21,
DOI 10.1007/978-3-319-64084-6_13

The paper is restricted to the comparison of only two types of rail transport - underground train and tramway transport, as strictly associated with urban transport. Although each extension of a rail transport system is linked to high investment costs, the paper does not include the issues of financial and economic analysis, for the advantage of functional aspects - offered by underground and tramway transport.

2 Transporting Capacity for the Underground and Tram Lines

The transporting capacity is particularly significant from the viewpoint of ensuring high quality of a transport system - also in the future, considering the different developmental scenarios for the city. Transporting capacity of a single line of urban transport depends on the size of the rolling stock and the frequency of runs of vehicles in service on the line. The size of the rolling stock is typically associated with the number of places offered in an underground train or a tram. Manufacturers of underground trains typically provide the limit occupancy, based on the occupancy of standing area of 7.0 [pers./m^2] (e.g. [4, 5]), that is under the conditions of extreme congestion. Tram manufacturers typically use occupancies calculated at 4.0 [pers./m^2] (e.g. [6]), or 5.0 [pers./m^2] (e.g. [7]), which is in line with the data applicable in the given city on the occupancy standards in public transport. This translates into the offered level of travel comfort, which is discussed (among others) in [8]. The distribution of seats in a vehicle is also variable, which results in differences of the numbers of offered places, even in trains with the same length, provided by the same manufacturers to different cities.

In practice, occupancy of standing area on the level of 4.0 [pers./m^2] is used. The frequency of runs, expressed by the number of trains per hour - may refer to both a single line (as is the case for underground) or bundle of lines going on the same route for its considerable stretch, which often takes place for tram lines, and results from the need to improve the directness of travel. In practice, the frequencies of the busiest sections of underground and tram networks do not exceed 40 [T/h]. The transporting capacity of a line or a line bundle, based on the data on the size of the rolling stock and frequency of runs can be calculated from the following formula:

$$\mathrm{C} = C_V \cdot f \qquad (1)$$

where:

C - transporting capacity of a line (line bundle) of public transport [ps./h],

C_v - offered volume of a public transport vehicle [ps./T] at the assumed comfort level for passenger travel,

f - frequency of travel of public transport vehicles [T/h].

Table 1 presents approximate offered number of places in one hour of operation, on hypothetical underground lines and tram lines, serviced by typical short, medium and long trains. In the period of one hour, the bundle of tram lines operated by long (large-volume) trams are capable of transporting up to over 10.5 thousand passengers in one direction. In comparison, currently in Krakow the highest hourly passenger

Table 1. Comparison of transporting capacity (in thousands persons per 1 h in one direction, occupancy of standing area of 4.0 [pers./m2]) of underground lines and tram lines depending on the interval and the size of the rolling stock.

Kind of vehicle	Size of vehicle (length [m])	Interval on public transport line [min]							
		1,5	2	3	4	5	6	7,5	10
Underground train	Short (52)	13.5	10.1	6.7	5.1	4.0	3.4	2.7	2.0
	Medium (100)	30.4	22.8	15.2	11.4	9.1	7.6	6.1	4.6
	Long (133)	42.0	31.5	21.0	15.7	12.6	10.5	8.4	6.3
Tramway	Short (26)	6.6	4.9	3.3	2.5	2.0	1.6	1.3	1.0
	Medium (32)	7.8	5.9	3.9	2.9	2.3	2.0	1.6	1.2
	Long (45)	10.6	8.0	5.3	4.0	3.2	2.7	2.1	1.6

streams do not exceed 6.0 thousand passengers [9], although the increase in the number of citizens and the traffic will influence the growth in the most significant urban routes. Underground train is far more efficient in this field, as it can handle streams reaching even 42.0 thousand passengers in one direction.

3 Underground Train and Tramway Travel Time

The second highly significant aspect of the rail transport system operation is the travel time, which has the greatest impact on the passenger selection of means of transport [10, 11], both between the individual and public transport, as well as the internal selections within the groups of different subsystems of the public transport, including underground, railway, trams, busses and trolleybuses. It also constitutes the basis for the modeling of division of transport tasks [12] in simulations of transport systems.

Each urban public transport travel can be divided into the following stages:

- time of reaching the stop - time, measured from the moment when a traveler leaves the mean of transport and reaches the stop, the time should take into account the real rout to the stop,
- waiting time for the public transport vehicle - from the moment of appearing at a stop to the moment of departure from the stop with an public transport vehicle,
- time of running with a public transport vehicle - from the moment of departure from the stop to the time of leaving the vehicle at the final stop,
- time of leaving the stop to the travel destination - time in which a traveler leaves the vehicle at the stop to the moment of reaching his travel destination,
- optionally - if during the travel a switch of the public transport line takes place, the time of transfer must be included, and it is measured from the moment of leaving one public transport vehicle in a transfer node (stop) to the time of departure from the node (stop) in a vehicle of another line.

The subsequent sections present an analysis of individual components of travel in underground train or tram.

3.1 Time of Reaching (Leaving) to (from) the Stop

The speed of a pedestrian in free movement depends, i.a., on his motor characters, motivation for the travel, knowledge of the timetable [13], frequency of vehicle runs and the state and safety of pedestrian infrastructure [14]. It typically lies in the range from 3.0 to 5.0 [km/h]. In the modeling of pedestrian traffic, the average value of 4.0 [km/h] is most commonly used, as it overs variable speeds at different sections of the route and also including the slowing down of pedestrians and their stops at junctions. Taken into account are also the natural distances between the conventional travel sources or destinations and the closes public transport stops, which are easy to reach (or leave). In practice, distances of reaching longer than 500 [m] are very rare and typically apply to peripheral or suburban areas [15]. Currently, actions are taken to ensure even low frequency bus lines, also in the areas of scattered housing. If reaching a stop requires facing a level difference - which is present for using tunnels - the mean time of overcoming the elevation difference should be added to the time of reaching and leaving a stop. Based on own study, it was established, that the time of overcoming elevations ranges between 50 [s] up to even 120 [s] in the case of deep underground stations. The average time of overcoming elevations can be also by physical dependencies. Considering the 30 or 35 degree slope of the escalator and the standard escalator speed of 0.5 [m/s], just overcome 8 [m] difference of levels (shallow underground or tramway tunnel) takes - respectively: 32 or 28 [s]. In case of 15 [m] deep tunnel it will be - 60 and 52 [s]. One should take into account the waiting time for the possibility of using the stairs, which depends on pedestrian traffic volume and the stair width and access to a vehicle entry point (escalators are generally located at the ends of platforms). It is assumed average waiting time of 13 [s] in both directions - down and up (from own, unpublished studies) for the mean conditions. Table 2 presents a list of mean times of reaching and leaving stops, depending on the length of a section, calculated along pedestrian paths.

Table 2. Average time of pedestrian reaching (leaving) from the stop depending on the length of the section.

Length [m]		100	200	300	400	500	600	700	800
Reaching (leaving) time [min]	On the ground tramway stop	1.5	3.0	4.5	6.0	7.5	9.0	10.5	12.0
	Underground stop (shallow tunnel)	2.4	3.9	5.4	6.9	8.4	9.9	11.4	12.9
	Underground stop (deep tunnel)	2.9	4.4	5.9	7.4	8.9	10.4	11.9	13.4

3.2 Waiting Time for a Public Transport Vehicle

The passengers' waiting time at a stop is directly linked to the length of the interval between subsequent urban transport vehicles. In general, in case of short intervals, almost a complete lack of relationship between the appearances of a passenger and the planned moment of departure from the stop is observed [16]. Similar situation takes place for tram line bundles, especially when a transit pass is not assigned to a route instead of a line. However, on the lines with low frequencies, appearances of passengers within 5 min to the moment of vehicle departing from the stop (according to time table) are predominant (81% of appearances, [17]) and the highest intensity can be observed within the 2-minute period prior the planned departure.

3.3 Public Transport Vehicle Travel Time

The travel time, including the time of running subsequent sections and the stopping time at subsequent stops [18] - is generally understood as the most important component of the travel time in public transport. This is because the component explains the movement at the greatest distance on the scale of the entire travel. The vehicle travel time depends on a variety of external factors, including many of a disturbing nature. Of high significance is here the state of used public transport infrastructure, including the presence and the range of separation of the rail vehicle traffic from the traffic of other vehicles and location of stops and stations. Attaining high travel speeds is possible when using an efficient separation of tracks and ensuring sufficiently large distances between stops, to enable attaining of a high speed. The highest level of track separation from the space available to other traffic users is - of course - ensured by underground transport.

An estimation for the average running time for all kinds of sections (underground, on the ground) can be carried out by a physical model, based on calculations of acceleration and deceleration times and running time with constant speed, can be used:

$$t_r = \frac{L}{v_{\max}} + \frac{v_{\max}}{2}\left(\frac{1}{a} + \frac{1}{b}\right) \tag{2}$$

where:

t_r - average running time for a section between stops [s],
L - length of a section [m],
v_{max} - maximum speed on the section [m/s],
a - acceleration rate [m/s^2],
b - deceleration rate [m/s^2].

This kind of model is very useful especially for underground train systems, where an automatic system control is implemented. But, it is not always efficient in description the movement of man-driven public transport vehicles, like tramways. For example - in Krakow exists three stop-to-stop tunnel sections for tramways with speed limits of 40 [km/h] - and the speeds achieved from measurements and from the model significantly differ. The model can be saved using a conversion coefficient β:

$$t_r = \beta \cdot \frac{L}{v_{\max}} + \frac{v_{\max}}{2}\left(\frac{1}{a} + \frac{1}{b}\right) \tag{3}$$

In the case of the mentioned tunnel sections in Krakow, it is 1.15. In such situations, regression models can work much better. That is why the empirical models based on the results of running time measurements in Krakow were undertaken. They were built on the basis of measurements' results carried out on 52 sections (length: 190–1065 [m], speed limit: 40–50 [km/h]), where the trams are moving along separate tracks. The research was conducted using manual GPS receivers. A significant database of results was obtained - some sections were measured after over 40 times. On this basis, the multiple regression model of the tram running time was built:

$$t_r = 0.093 \cdot L + 7.78 \cdot N_s + 8.27 \cdot N_z \tag{4}$$

where:

N_s - number of junctions at the section,

N_z - number of tramway switches at the tramway unions on the section between two stops.

However, in the case of sections without junctions and tramway switches – the model returns too short travel times, difficult to obtain during real operation of the tramway line. So, it was decided to build a model only for sections without junctions and tramway switches. A simple regression model was conducted, where section running time depends only on the length of the section:

$$t_r = \frac{1}{0.102 - 0.013\ \ln(L)} \tag{5}$$

The model is characterized by particularly high determination coefficient - the model explains 90.3% of the variability in section running time. The result of the ANOVA test gives a statistically significant relationship between running time and the length of the section, at the 95.0% confidence level. Both model components (intercept and slope) are also statistically significant. Figure 1 shows the comparison of obtained models (physical models with maximum speed 40 and 50 [km/h], multiple regression model and simple regression model) for the range of section length between 300 and 700 [m], what describes the majority of stop-to-stop distances on tram networks in city centers. Whereas Fig. 2 compares the speed of travel according to the length of the sections.

Three of the models presented are susceptible to changes in the length of the section between following stops. Only a multiple regression model offers constant speed - which reduces its usability. In other cases, the length of the section is essential for the running speed on the section with separated tramway track.

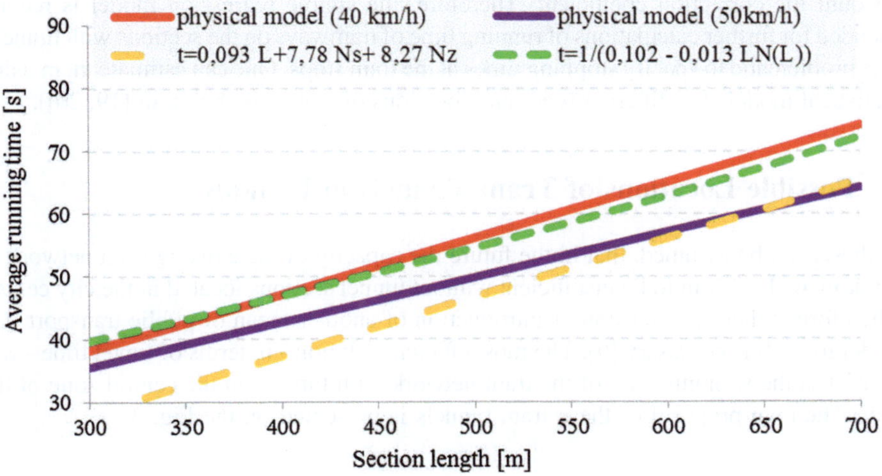

Fig. 1. Relationship between average travel time of tram line section and the length of the section, for considered models

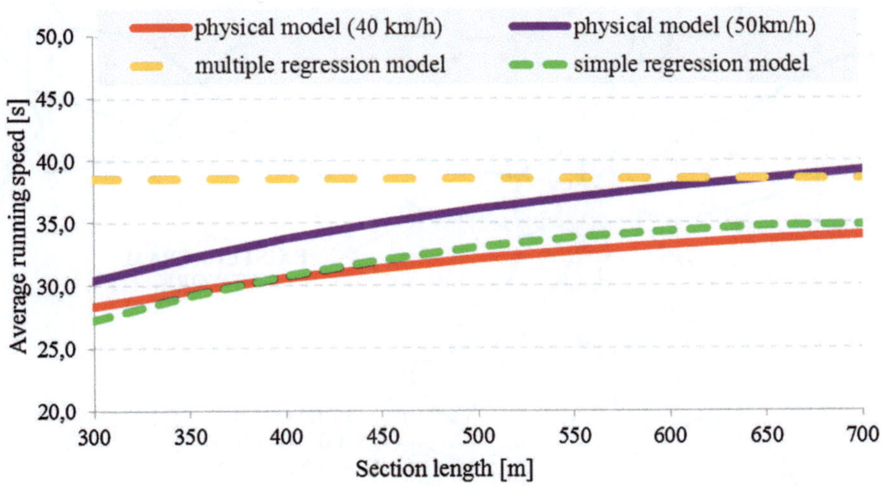

Fig. 2. Comparison of running speeds, for considered models

One can also see a high convergence of the results obtained using the physical model with a maximum speed of 40 [km/h] and a simple regression model. This means, that in Krakow, on the sections with separated tracks (without junctions and tramway switches) - in spite of the most frequent limits of 50 [km/h] - tramways achieve relatively slower running speeds. This is due to the high intensity of tram traffic on these sections.

As already mentioned, the only three tunnel sections in Krakow, where there are limits of 40 [km/h], running time obtained from the physical model must take into account the correction coefficient. Therefore, the simple regression model is recommended for further calculations of running time of tramways on the sections with tunnels. It is problematic to specify stopping times at the tram stops. One can estimate them using statistical models for different types and locations of stops (as shown in [19, 20]).

4 Possible Locations of Tram Tunnels in Krakow

A thesis can be assumed, that in the future the capacity of the existing tram network in Krakow will turn out to be insufficient without tunnel sections located in the city center. Therefore, either its extension or introduction of another mean of public transport (i.e. underground) is necessary [9]. The most efficient solution - in terms of travel time - will consist in the strengthening of the tram network with tunnels in the central zone of the city. The own proposal of these tram tunnels is presented on the Fig. 3.

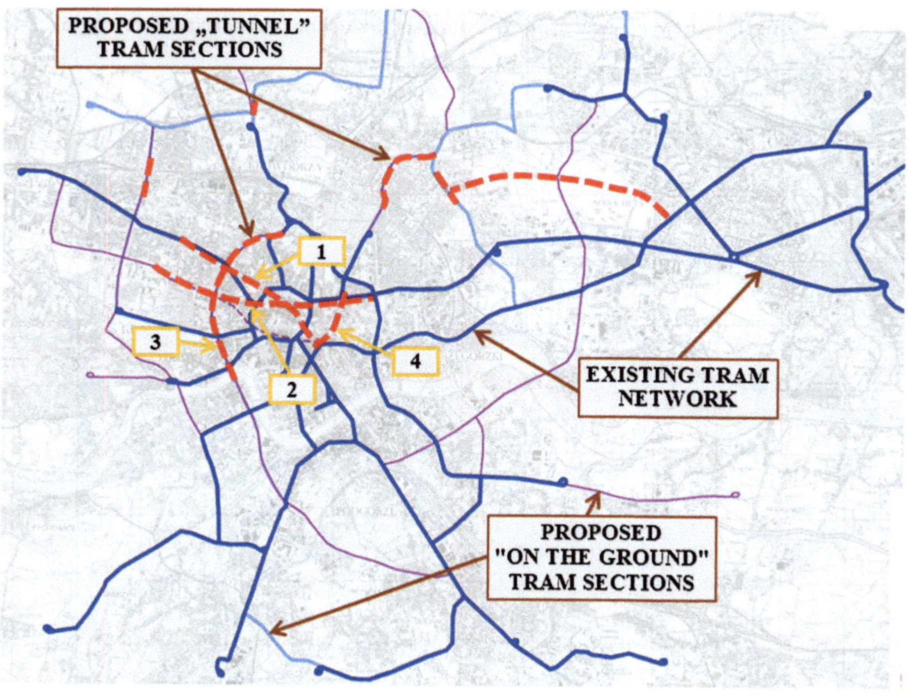

Fig. 3. Proposed tram routes against the existing network

Construction of new tunnel tram routes (among others) have been proposed:

- section 1: from stop "Kawiory" to the public transport node "Rondo Mogilskie",
- section 2: from stop "Urzędnicza" to stop "Hala Targowa",

- section 3: from the public transport node "Nowy Kleparz" to the public transport node "Centrum Kongresowe" (along the second city ring road),
- section 4: from Rakowicka street to the public transport node "Hala Targowa".

It will facilitate a significant increase for the efficiency of the whole tram system, yet without a significant extension of the distance between stops. Thus, the somewhat forgotten idea of pre-underground will be introduced. Only the section number 2 appears in the official city documents [9]. But, the discussion on the extension of the transport system of Krakow has not yet been completed and in the future the number and length of planned tram tunnels may increase.

5 Expected Effects of Tramway Tunnels in City Center of Krakow

Expected effects of tramway tunnels were estimated for tunnels number 1 and 2, located in close city center of Krakow. Figure 4 shows the proposed location of tram stops at these tunnel tramway lines.

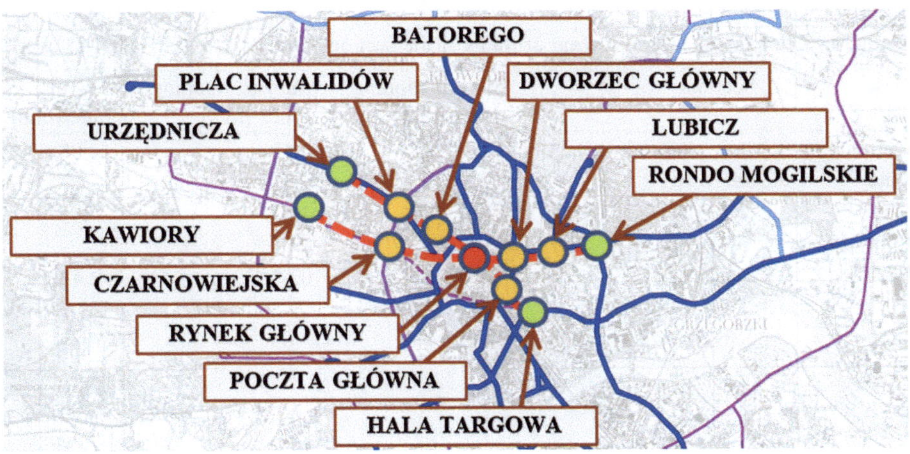

Fig. 4. Tram stops at the proposed tunnel routes in close city center of Krakow

Opening new tram tunnels will effect by partially transfer existing lines into tunnels. Initial analyzes show, that tram traffic volumes in both tunnels will achieve 18 vehicles per hour (average interval: 2.33 [min], one line at 12 T/h and one line at 6 T/h - in every tunnel). General information about these tunnels are shown in Table 3.

New tunnel tram routes will shorten the travel time to the city center of Krakow. In the existing state - the time of access from stops located on the I ring to the Main Market Square is an average of 4 min. Despite the need to overcome the difference in levels, it will shorten by an average of 1.3 min. New tunnels will improve also access to the other trip destinations within the strict city center.

Table 3. General information about proposed tramway tunnels.

Section	Length [m]	Number of underground stops [-]	Frequency [T/h]	Capacity (one dir.) [pers./h]
“Urzędnicza” - “Hala Targowa”	2660	4	18	4800
“Kawiory” - “Rondo Mogilskie”	3700	4	18	4800

However, a much more pronounced effect will occur in the case of travel time to the city center. Comparison of current and future travel times is shown in Table 4 (section: “Urzędnicza” - “Hala Targowa”) and Table 5 (section: “Kawiory” - “Rondo Mogilskie”).

One can see, that new tramway tunnels decrease the travel time on discussed sections. In both cases - this is about 50 [%] effect for passengers. It should also be added, that the capacity of the tramway network in the city center will increase significantly. Obtained results can be used in the macro-stimulation analysis of the transportation system of Krakow, as initial conditions.

Table 4. Comparison of travel time of trams at the section “Urzędnicza” - “Hala Targowa”.

Case	Section	Sum of running times [min]	Sum of stopping times [min]	Travel time [min]
Current state	“Urzędnicza” - “Hala Targowa”	12.88	2.10	14.98
With tramway tunnel	“Urzędnicza” - “Main Market Sq.”	3.01	1.43	4.45
	“Main Market Sq.” - “Hala Targowa”	1.78	0.62	2.40
	“Urzędnicza” - “Hala Targowa”	4.79	2.05	6.84

Table 5. Comparison of travel time of trams at the section “Kawiory” - “Rondo Mogilskie”.

Case	Section	Sum of running times [min]	Sum of stopping times [min]	Travel time [min]
Current state	“Kawiory” - “Rondo Mogilskie”	14.50	2.42	16.92
With tramway tunnel	“Kawiory” - “Main Market Sq.”	1.12	0.50	1.62
	“Main Market Sq.” - “Rondo Mogilskie”	3.01	0.92	3.93
	“Kawiory” - “Rondo Mogilskie”	6.75	1.83	8.58

6 Conclusions

The paper discusses the transporting capacity and travel times of different rail transport systems. A discussion has been conducted on the influence of underground tramway transport onto accessibility of the city center of Krakow and the travel time of tramways in daily operations. As a result of the discussion, it was determined, that a particularly high functional efficiency of the Krakow's transportation system should be ensured by new tram connections which will utilize tunnels.

References

1. Feng, X., Mao, B., Feng, X., Feng, J.: Study on the maximum operation speeds of metro trains for energy saving as well as transport efficiency improvement. Energy **36**(11), 6577–6582 (2011)
2. Hensher, D., Button, K.J.: Handbooks in Transport. Emerald, United Kingdom (2007)
3. Sierpiński, G.: Theoretical model and activities to change the modal split of traffic. In: Mikulski, J. (ed.) Telematics in the Transport Environment. CCIS, vol. 329, pp. 45–51. Springer, Heidelberg (2012)
4. Siemens Website, http://www.siemens.com/press/innotrans2012
5. Transport for London: Rolling Stock Data Sheet. Transport for London, London (2007)
6. Bombardier Website, http://www.bombardier.com/en/transportation/products-services/rail-vehicles.html
7. Pesa Website. http://pesa.pl/press-room/nowosci/111=swing
8. Redman, L., Friman, M., Gärling, T., Hartig, T.: Quality attributes of public transport that attract car users: a research review. Transp. Policy **25**, 119–127 (2013)
9. Transport Engineering Office: Study of the Development of Cracow's Transport System Including the Metro Construction. Transport Engineering Office, Poznań (2015)
10. Santos, G., Maoh, H., Potoglou, D., Brunn, T.: Factors influencing modal split of commuting journeys in medium-size european cities. J. Transp. Geogr. **30**, 127–137 (2013)
11. Sierpiński, G.: Changes of the modal split of traffic in Europe. Arch. Trans. Syst. Telemat. **6**, 45–48 (2013)
12. Jacyna, M., Wasiak, M., Lewczuk, K., Kłodawski, M.: Simulation model of transport system of poland as a tool for developing sustainable transport. Arch. Trans. **31**, 23–35 (2014)
13. Mikłasz, M., Olszewski, P., Nowosielski, A., Kawka, G.: Pedestrian traffic distribution analysis using face recognition technology. In: Mikulski, J. (ed.) Activities of Transport Telematics. CCIS, vol. 395, pp. 303–312. Springer, Heidelberg (2013)
14. Galanis, A., Eliou, N.: Evaluation of the pedestrian infrastructure using walk ability indicators. WSEAS Trans. Environ. Develop. **7**, 385–394 (2011)
15. Murray, A.T.: Strategic analysis of public transport coverage. Socio-Econ. Plan. Sci. **35**, 175–188 (2001)
16. Karoń, G., Macioszek, E., Sobota, A.: Selected problems of transport network modelling of upper-silesian agglomeration (In Poland). In: Proceedings of the 6th International Scientific Conference Transbaltica, pp. 103–108. Vilnius Gediminas Technical University, Vilnius (2009)
17. Rudnicki, A.: Quality in Urban Public Transportation. SITK, Krakow (1990)

18. Bauer, M.: Application of GPS technology to evaluate the quality of public transport. Acta Technica Jaurinensis **6**(3), 11–23 (2013)
19. Dueker, K.J., Kimpel, T.J., Strathman, J.G., Callas, S.: Determinants of bus dwell time. J. Public Trans. **7**(1), 21–40 (2004)
20. Bauer, M., Richter, M., Weiß, H.: Simulation Model of Tram Route Operation. Bauhaus-Universität Weimar, Weimar (2010)

Using GT Planner to Improve the Functioning of Public Transport

Ireneusz Celiński(✉)

Faculty of Transport, Silesian University of Technology, Katowice, Poland
ireneusz.celinski@polsl.pl

Abstract. An increase of the share of public transport in the total number of travels within the transport network is a key aspect in the efforts aimed at achieving sustainable transport development. It is impossible unless the functioning of this branch of transport is improved, including particularly greater availability of public transport lines. The development of public transport IT systems was supposed to have improved at least the identification of demand for transport services. Practice has shown, however, that it is by no means as simple as that. The article presents a general idea of particular method of identifying the demand for public transport services. It is a method that operates outside of the public transport (PuT) system's roads and means of transport. It also makes it possible to stimulate PuT use among people using private transport (PrT) on a day-to-day basis. To illustrate this, the article uses the Green Travelling Planner (GT Planner) functionality.

Keywords: Public transport · Green travelling planner · Trip planner

1 Introduction

A key limitation in the transport network is the traffic carrying capacity of its cross-sections [1–3]. The existing modal split of transport tasks has led to the total or progressing depletion of traffic carrying capacity in numerous cross-sections of the road network, with substantial reserves of that capacity left, on the other hand, in other transport branches (for instance rail transport). Within the road network, the modal split between public and private transport may also be modified, with a consequent increase of the capacity reserve [4–10]. This entails, at the same time, a significant improvement of traffic flow. An essential problem in this respect is how to encourage private transport (PrT) users to use public transport (PuT), i.e. trains, trams and buses [11]. This is impossible unless an appropriate supply is provided, and this in turn requires existing and suppressed demand to be identified in this field of transport services [12, 13].

The development of public transport IT systems is supposed to help in the identification of existing demand by examining the streams of travelling persons in public transport (analyses of occupancy and of travel routes). Associating a travel registered in a specific means of public transport with the travelling person's data (place of residence) makes it possible to reconstruct nearly completely the existing traffic structure with regard to public transport. The main issue when it comes to the accurate reconstruction of that traffic structure concerns the path of departure from the public transport

E. Macioszek and G. Sierpiński (eds.), *Recent Advances in Traffic Engineering for Transport Networks and Systems*, Lecture Notes in Networks and Systems 21,
DOI 10.1007/978-3-319-64084-6_14

line access point to the transfer absorption point. For the dominant types of travelling motivations, i.e. H-W-H and E-H-E, the problem can be solved using generally available data, for instance those available to tax authorities.

Practice shows, however, that the use of public transport IT systems (city cards, occupancy meters, Passenger Information Systems) encounters unexpected problems [14]. For instance, travelling persons using electronic urban transport cards do not accept the way in which they have to use them (the card needs to be held against the card reader many times also when the passenger is leaving the vehicle) [15]. The vast majority of passengers continue to travel without such a card, including people who use public transport services regularly. What is most important, however, such systems currently do not monitor the transfers of persons who use private transport. Therefore, the demand parameters are not identified to a sufficient extent in either group, namely that of existing and that of potential public transport users (respectively EPuT and PPuT). In the PPuT group (i.e. travelling persons who may switch to public transport for their transfers), these needs are essentially not identified at all [4, 6, 8, 9]. In the group of private transport users (PrT), this identification is limited to indirect studies carried out as part of the study of transport behaviour patterns.

As a consequence of this insufficiency of knowledge about the existing and potential demand for public transport, it is impossible for the operators to adjust their offering to the existing and potential needs. Therefore, the evidence of unsustainable development of transport observed on a daily basis falls to a significant extent within the competence of public transport organisers. The problems that need to be enumerated here include poor availability of public transport, both in space and in time, as well as the inappropriate organisation of public transport lines, both in the modal and in the comodal aspect.

This article presents an IT tool referred to as Green Travelling Planner (GT Planner), which may lead to a significant reduction of the barriers to sustainable development of transport listed above [16]. This is supposed to be achieved by identifying the transport-related needs in all user groups.

2 GT Planner

GT Planner is a travel planner focusing specifically on multimodal and eco-friendly transfers. Similarly to most travel planners, this tool determines the connection (transfer path) between the specific points of departure and arrival. What makes this particular planner different is the optimisation of the transfer path within the transport network, not only using the shortest or most economical route, but also using multimodal connections and marking the transfer with emission indicators. The task of GT Planner consists in encouraging its users to choose eco-friendly forms of travel within the transport network covered by the tool. An example of how a transfer path is determined within the transport network of the Upper Silesian conurbation using GT Planner is shown in Fig. 1 [16].

As the user interacts with the GT Planner interface, parameters may be recorded related to the requested transfer with regard to many different variables. Apart from the departure and arrival points of the travel, the following are also recorded: a set of

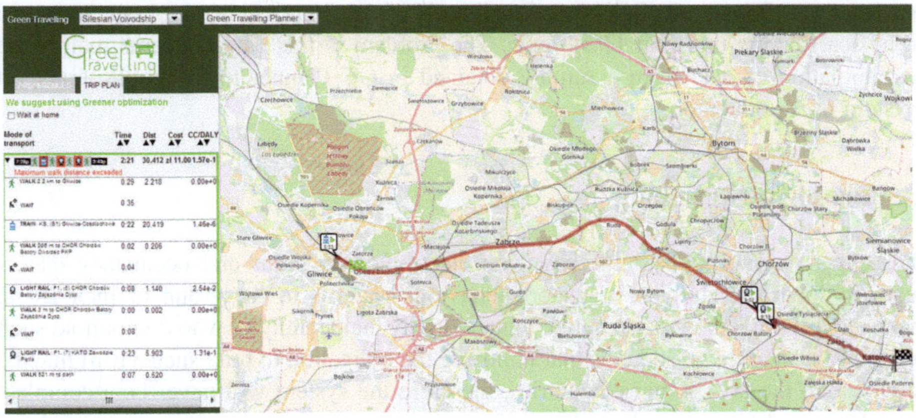

Fig. 1. GT Planner, main window

intermediate points, the travel motivation, the means of transport used, and optimisation preferences (eco-friendly or multimodal travel, the quickest, shortest, cheapest travel, etc.).

What makes GT Planner differ from other public transport IT systems is the fact that the transport behaviour patterns are analysed outside of it, and above all outside of the means of transport related to the transfer, to be used by the planner user. Consequently, the data from the planner constitute a certain forward-looking estimate, not a concrete value. The use of the GTFS standard in GT Planner leads at the same time to a very detailed description of the planned transfers using public transport with their simultaneous positioning in space and time within the transport system being analysed (positioning in the geocentric system using GPS) [17, 18]. Apart from that, GT Planner makes it possible to study not only transport behaviour patterns (as it happens in the case of public transport IT systems), but also transport preferences (unrealised behaviour patterns). In GT Planner, these are travel requests that will not be fulfilled for various reasons. They include cases when the planner is unable to indicate a connection or when the connection indicated does not suit the user (and is abandoned). Consequently, a travel planner of the GT Planner type has, in certain respects, a much more extensive functional scope than the systems used to monitor transport behaviour patterns relying on data from city cards or occupancy meters, or analyses of streams of travelling persons on specific sections. In practice, in existing traffic monitoring systems for public transport, the part of the demand that is studied is the part balanced out to a substantial extent by the supply. It is therefore essentially a partial study, which does not solve the problem in the broader context.

Currently, users of travel planners in the area of the Upper Silesian conurbation account for around 6% of the population. According to US studies conducted in California in the early 2000s, the potential number of planner users may reach 42% of the population [19–21]. These figures suggest that travel planners may constitute a source of data concerning demand for public transport in the future, not inferior by any means to the currently used passenger monitoring systems. On the other hand, tools such as GT Planner include functionalities that are impossible to obtain in actual (physical) public transport IT systems. The problem in this case concerns mainly the extrapolation of data from a sample obtained in GT Planner to the public transport system user population.

3 Problem Statement

Passenger monitoring systems based on electronic city cards and occupancy meters narrow down the study of transport behaviour patterns to the group of their users, ignoring passengers who use public transport only incidentally or who do not hold such cards. What should be added is that imposing the obligation to hold such a card may be contrary to the law in many countries, therefore the system will never be a complete one. Systems based on such cards are affected by an error resulting from the procedures adopted for its operation, as well as from system errors and failures. Furthermore, they are incapable of analysing the potential demand, since they only monitor the existing one [12, 13]. This makes data obtained from tools such as GT Planner more useful in the process of improving the sustainable development of transport. The global demand (*TD*) for services offered by public transport operators within a certain limited space (the area of analysis), after its delimitation into (*i* * *j*) spatial regimes, may be expressed with the following equation:

$$TD \approx \sum_{k=1}^{TM} \left(\sum \sum D_{ij}^{E} + \sum \sum D_{ij}^{P} \right) + \sum \sum D_{ij}^{PC} \tag{1}$$

where:

D_{ij}^{E} - planned demand in spatial regime ij [Number of travels planned using pub lic transport],
D_{ij}^{P} - potential demand in spatial regime ij [Number of rejected travels, sup pressed demand],
D_{ij}^{PC} - planned demand by passenger in spatial regime ij [Number of travels planned using a private vehicle, personal car],
ij - spatial regime indices (horizontal and vertical),
TM - number of different means of public transport within the area of analysis.

As opposed to other public transport IT systems, a travel planner such as GT Planner is already capable of analysing all the components included in Eq. (1). The analysis may be based on samples whose size exceeds those used in studies of transport behaviour patterns in urban areas which used to be considered as representative until now (1% to 5%) [19]. Planned demand D_{ij}^{E} is determined on the basis of the requests addressed to GT Planner in spatial regime which public transport was selected as the means of transport. Hidden demand D_{ij}^{P} is determined on the basis of the number of enquiries for which no transfer path was determined (limited PuT offering) or the means of transport was changed to a non-public one. The demand fulfilled with private transport D_{ij}^{PC} is determined directly on the basis of the number of requests for a transfer path using private means of transport (motor car, including electric cars; bicycle, including city bikes; longer transfers on foot). Consequently, a set of at least three demand quantities (indices) is determined for each spatial regime within the area of analysis. When public transport is used that involves more than one means of transport, the number of indices is multiplied by the respective number (Fig. 2).

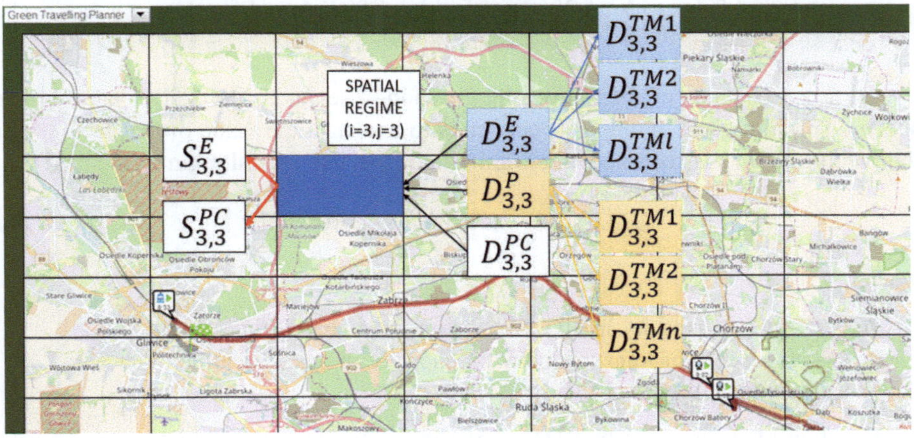

Fig. 2. GT Planner, delimitation and spatial regime factor

In response to the demand reported in the specific spatial regime (in this case: actual and hidden aka suppressed demand), the means of public transport at the level of S_{ij}^{E} needs to be organised in order to improve the sustainable development of transport (an appropriate number of seats and/or standing places needs to be guaranteed). It is worth pointing out in this context that the supply in the specific spatial regime for transfers using private transport S_{ij}^{PC} may be determined at the level of the total traffic carrying capacity of the road network cross-sections or as the transport capacity of the surface area corresponding to the respective regime. If one of the operating criteria of a transport system organisation is its sustainable development, it should be possible to describe the public transport system supply within the spatial regime with the following relation:

$$S_{ij}^{E} \geq \sum_{k=1}^{TM_{ij}} \left(D_{ij}^{E} + D_{ij}^{P} \right) + STDF * D_{ij}^{PC} \quad (2)$$

where:

S_{ij}^{E}—number of seats/places in PuT in spatial regime *ij*,

STDF—sustainable transport development factor (transfer rate between PrT and PuT).

When determining the STDF value (percentage share of traffic that moved from private transport to public transport), the required supply of seats/standing places on means of public transport is determined in each spatial regime within the area of analysis. One needs to bear in mind that due to the fact that traffic varies in time and space, this factor should be different and vary in time in each spatial regime, just like the demand values recorded using GT Planner:

$$S_{ij}^{E}(t) \geq \sum_{k=1}^{TMij} \left(D_{ij}^{E}(t) + D_{ij}^{P}(t) \right) + STDF_{ij}(t) * D_{ij}^{PC}(t) \tag{3}$$

As a consequence of the above, the question arises as to the STDF value in the specific spatial regime. Intuitively, one tends to associate the value of that factor with the parameters of the road network within the respective spatial regime:

$$STDF_{ij}(t) \propto \quad \Delta RC(\Delta t) \quad \| \Delta RCC(\Delta t) \tag{4}$$

where:

RC - reserve of capacity in all cross-section in spatial regime [Veh./min],
RCC - or/ reserve of transport network capacity in spatial regime [Veh./km],
Δt - measurement interval [Min].

Equation (4) means that the$STDF$sustainability factor in the spatial regime should be proportional to the value of capacity reserve changes and/or of transport network capacity changes within the spatial regime in the analysed period.

Application of the method for determining demand in spatial regimes of the road network should result in a map defining the demand reported in all of its regimes. An example of such a map is shown in Fig. 3.

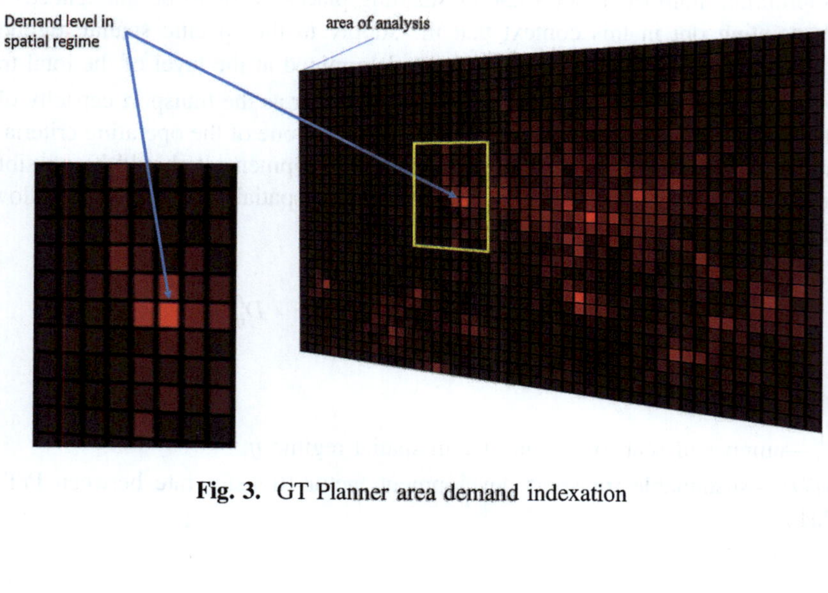

Fig. 3. GT Planner area demand indexation

The demand determined using Eq. (1) concerns road network users who have access to the World Wide Web and use travel planning systems (for example GT Planner). In order to create the map presented in Fig. 3, the value of that demand in each spatial regime should be increased by the value of the relevant coefficient:

$$SRC_{ij}(t) = \frac{NT_{ij}^{ITS}(t)}{TD_{ij}^{GT}(t)} \tag{5}$$

where:

$SRC_{ij}(t)$ - spatial regime coefficient,
$NT_{ij}^{ITS}(t)$ - number of transfers observed in the given regime (counters, loops etc.),
$TD_{ij}^{GT}(t)$ - planned demand calculated in GT Planner in the given regime [Number of planned and rejected travels].

In order to calculate the value of coefficient (5), the number of travels observed in the spatial regime should be divided by the number of travels planned in the given spatial regime. The number of travels observed can be measured using induction loops, occupancy meters, etc. (ITS, speaking more broadly). Discussion on the method of extrapolation of the demand obtained in GT Planner to the populations exceeds the framework of this article. Knowing what the total demand for the population in each spatial regime is (including suppressed demand), one can compare it to the existing supply (the number of seats/places offered on the means of transport):

$$B_{ij}(t) = TD_{ij}^{G}(t) - S_{ij}^{G}(t) \tag{6}$$

where:
$B_{ij}(t)$ - temporary balance in the spatial regime [Seats/places],
G - index referring to the total demand or total supply in the spatial regime.

Both the positive and the negative values of Eq. (6) constitute grounds for public transport organisers to undertake legitimate actions. Positive values proving unsatisfied demand should constitute the basis for increasing vehicle capacity within the given spatial regime, increasing running frequency or reorganising the public transport line routing. Line routing reorganisation is justified inasmuch as seat/place reserves are available on means of public transport in neighbouring regimes.

Using GT Planner functionalities makes it possible to analyse demand parameters also taking into account the adjacent spatial regimes. This is significant due to the possibility of balancing seats/places on means of public transport between neighbouring spatial regimes (small adjustments to the line routing). Each of the demand values included in Eq. (1) may be expressed with the following sum:

$$D_{ij}^{*}(t) = DN_{ij}^{*} + DW_{ij}^{*} + DE_{ij}^{*} + DS_{ij}^{*} \tag{7}$$

where:

* - existing or suppressed demand [Seats/places],
N,W,E,S - respectively: routes of transfers towards the northern, western, eastern and southern boundaries of the spatial regime.

Equation (7) may be used to calculate the balance of supply and demand for public transport services in the spatial regime with regard to the individual boundaries of that regime. When this approach is used, the layout of public transport lines will consist in searching for a transfer path between spatial regimes set out in a three-dimensional data table containing the values of the balance expressed by Eq. (6). Discrete optimisation methods can be used for this purpose. Various cases of public transport line routing optimisation using this method will be presented in a subsequent publication.

4 Conclusions

The general framework of the method to improve the functioning of public transport proposed in this article does not require any substantial expenditures. Above all, it requires dissemination and education with regard to the use of a planner of the GT Planner type [11]. US studies indicate that nearly every other transport network user may contribute significant information to an extent that is essential for the improvement of public transport operation. Additional knowledge of the suppressed demand value in each regime will make it possible to improve public transport organisation [12, 13]. Travels may be moved between public transport and private transport (justified) when the limitations are known in the individual spatial regimes. If the transport network is delimited into spatial regimes with small surface areas, this will enable the organisation of that transport to be modified in a relatively precise manner. This method requires building an specified IT tool for this purposes. Such a tool can be based on microsimulation software and GT Planner data.

Big changes in the modal split in spatial regime are associated with changes in road load. Adding heavy vehicles (buses, articulated buses) causes more intensive use of road pavement. This causes the process of sustainable development must be monitored using appropriate methods to road inspection [22–24].

Moving traffic from PrT to PuT also requires analysis for specific infrastructure solutions [25]. Such a process cannot be implemented without considering this aspect of road network. Increase the share of large vehicles must be monitored closely e.g. in terms of sections geometry (for example on the roundabouts).

The article does not take into account the impact of bicycle traffic on sustainable development [26–28]. This kind of traffic can be an essential ingredient of suppressed demand as well as can constitute transfer from PrT. This problem requires further research in this with GT Planner help.

The given method should be supported by transport network parameterisation using the GTAlg tool (part of GT Planner) [29]. This kind of transport network parameterization allows to justify adding the extra PuT vehicles in the strictest spatial regime (adding new bus stop with line correction).

Similarly, it is important to remember that changes in the number of new PuT vehicles entered into road network will affect other aspects of operation in it. These changes concern aspect of transport comodality [30].

Acknowledgements. The present research has been financed from the means of the National Centre for Research and Development as a part of the international project within the scope of ERA-NET Transport III Future Travelling Programme “A platform to analyze and foster the use of Green Travelling options (GREEN_TRAVELLING)”.

References

1. Transportation Research Board: Highway Capacity Manual, 6th edn. A Guide for Multimodal Mobility Analysis. Transport Research Board, Washington (2016)
2. May, A.D.: Performance Measures and Levels of Service in the Year 2000 Highway Capacity Manual. Final Report. NCHRP Project 3-55(4). Transport Research Board, Washington (1997)
3. Duranton, G., Turner, M.A.: The fundamental law of road congestion: evidence from US cities. Am. Econ. Rev. **101**, 2616–2652 (2011)
4. Sierpiński, G.: Distance and frequency of travels made with selected means of transport - a case study for the upper silesian conurbation (Poland). In: Sierpiński, G. (ed.) Intelligent Transport Systems and Travel Behaviour. AISC, vol. 505, pp. 75–85. Springer, Heidelberg (2017)
5. Huang, C., Chen, C., Wang, B., Dai, Y., Zhao, J., Wang, H.: Urban travel modal split and its impact on energy and environment. J. Highw. Transp. Res. **22**, 163–165 (2005)
6. Sierpiński, G.: Travel behaviour and alternative modes of transportation. In: Mikulski, J. (ed.) Modern Transport Telematics. CCIS, vol. 239, pp. 86–93. Springer, Heidelberg (2011)
7. Hensher, D.A., Button, K.J.: Handbook of Transport Modelling. Pergamon, Amsterdam (2007)
8. Sierpiński, G.: Model of incentives for changes of the modal split of traffic towards electric personal cars. In: Mikulski, J. (ed.) Telematics—Support for Transport. CCIS, vol. 471, pp. 450–460. Springer, Heidelberg (2014)
9. Sierpiński, G.: Theoretical model and activities to change the modal split of traffic. In: Mikulski, J. (ed.) Telematics in the Transport Environment. CCIS, vol. 329, pp. 45–51. Springer, Heidelberg (2012)
10. Yang, L., Guo, Z., Xiaoning, Z.: Cross-nested logit model for the joint choice of residential location, travel mode, and departure time. Habitat Int. **38**, 157–166 (2013)
11. Sierpiński, G., Staniek, M., Celiński, I.: New methods for pro-ecological travel behavior learning. In: 8th International Conference of Education, Research and Innovation, pp. 6926–6933. IATED, Seville (2015)
12. Szarata, A.: The simulation analysis of suppressed traffic. Adv. Transp. Stud. **29**, 35–44 (2013)
13. Szarata, A.: Modelowanie Symulacyjne Ruchu Wzbudzonego i Tłumionego. Transp. Miejski i Reg. **3**, 14–17 (2010)
14. Website, https://portal.kartaskup.pl
15. Website, http://www.dziennikzachodni.pl/serwisy/komunikacja/a/odbijanie-karty-skup-z-problemami-znowu,10113456/

16. Sierpiński, G.: Technologically advanced and responsible travel planning assisted by GT planner. In: Macioszek, E., Sierpiński, G. (eds.) Contemporary Challenges of Transport Systems and Traffic Engineering. LNNS, vol. 2, pp. 2367–3370. Springer, Heidelberg (2017)
17. Staniek, M., Sierpiński, G., Celiński, I., Czech, P., Barcik, J.: Implementation of the public transport timetable in GTFS standard. Logistyka **4**, 1507–1512 (2015)
18. Website, https://developers.google.com/transit/gtfs/reference/?csw=1
19. Celiński, I., Sierpiński, G., Staniek, M.: Sustainable development of the transport system through rationalization of transport tasks using a specialised travel planner. In: Dell'Acqua, G., Wegman, F. (eds.) Transport Infrastructure and Systems, pp. 1071–1080. Taylor and Francis, London (2017)
20. Gretzel, U.: Consumer generated content. trends and implications for branding. Rev. Tour. Res. vol. **4**(3), pp. 9–11 (2006)
21. Gretzel, U., Yoo, K., Purifot, M.: Online Travel Review Study, Role & Impact of Online Travel Reviews. Texas University, Texas (2007)
22. Staniek, M.: Road pavement condition as a determinant of travelling comfort. In: Sierpiński, G. (ed.) Intelligent Transport Systems and Travel Behaviour. AISC, vol. 505, pp. 99–107. Springer, Heidelberg (2017)
23. Sayers, M.W., Gillespie, T.D., Peterson, W.D.O.: Guidelines for Conducting and Calibrating Road Roughness Measurements. The International Bank for Reconstruction and Development, Washington (1986)
24. Staniek, M.: Stereo vision method application to road inspection. Baltic J. Road Bridge Eng. **12**, 38–47 (2017)
25. Macioszek, E.: Relationship between vehicle stream in the circular roadway of a one-lane roundabout and traffic volume on the roundabout at peak hour. In: Mikulski, J. (ed.) Telematics—Support for Transport. CCIS, vol. 471, pp. 110–119. Springer, Heidelberg (2014)
26. Macioszek, E., Sierpiński, G., Czapkowski, L.: Problems and issues with running the cycle traffic through the roundabouts. In: Mikulski, J. (ed.) Transport Systems Telematics. CCIS, vol. 104, pp. 107–114. Springer, Heidelberg (2010)
27. Schwartz, W.: Forecasting Bicycle and Pedestrian Demand. In: 68th ITE Annual Meeting and Exhibit, pp. 1–9. Institute of Transportation Engineers, Toronto (1998)
28. Macioszek, E., Sierpiński, G., Czapkowski, L.: Methods of modeling the bicycle traffic flows on the roundabouts. In: Mikulski, J. (ed.) Transport Systems Telematics. CCIS, vol. 104, pp. 115–124. Springer, Heidelberg (2010)
29. Celiński, I.: Transport network parameterisation using the GTAlg tool. In: Macioszek, E., Sierpiński, G. (eds.) Contemporary Challenges of Transport Systems and Traffic Engineering. LNNS, vol. 2, pp. 111–123. Springer, Heidelberg (2017)
30. Jacyna, M., Kłodawski, M.: Model of Transportation Network Development in Aspect of Transport Comodality. In: 21st International Conference on Systems Engineering, pp. 341–345. IEEE Press, Las Vegas (2011)

Development of the Bike-Sharing System on the Example of Polish Cities

Piotr Czech(✉), Katarzyna Turoń, and Grzegorz Sierpiński

Faculty of Transport, Silesian University of Technology, Katowice, Poland
{piotr.czech, katarzyna.turon, grzegorz.sierpinski}@polsl.pl

Abstract. The work was dedicated to the application of the concept of a system of urban public bicycle as an alternative means of transport in Polish cities. Authors presented operating systems on Polish territory, then referred to the usage statistics by individual customers self-service bike–sharing. They showed differences in the functioning and availability of systems and presenting in detail the principle of bike reservation by Wavelo system that is functioning in Cracow. The article also presents the advantages and disadvantages of the bike-sharing system. The aim of the study was to present the concept of self-service public bicycle rental system in Poland and its availability, popularity and prospects for future development.

Keywords: Bike-sharing system · Bike-sharing system in poland · Urban logistics · Urban mobility solutions · Sustainable development

1 Introduction

Searching for newer and newer ways to convince the urban population to use alternative forms of transportation is one of the objectives of sustainable transport [1–3]. There are a lot of studies related to specific territory or road solution type (see [4–6], among others). Sometimes information acquired directly from vehicles is very important also (see [7–9], among others). One of the forms of ecological, economical and both good for the health way of movement in urban transport is to travel by bike. Traveling with this mean of movement is not a new idea, but the concept of transport by bicycle without having the bike on the property is becoming more and more popular and appreciated by the inhabitants of Polish cities.

The aim of the study is to present the possibilities which are inherent to the concept of bike-sharing in Polish cities, as well as a reference to the availability, popularity of the system and the prospects for future development.

2 Bike-Sharing Systems Development

The idea of bike-sharing based on the possibility of self-service renting a bike and use it through the city wherein the system is implemented [10]. That concept allows the systems users have a real influence on the three basic principles of sustainable development like:

E. Macioszek and G. Sierpiński (eds.), *Recent Advances in Traffic Engineering for Transport Networks and Systems*, Lecture Notes in Networks and Systems 21,
DOI 10.1007/978-3-319-64084-6_15

- Environmental aspect - reducing the negative impact of conventional means of transport on the environment (emissions, vibrations),
- The economic aspect - increasing interest in cities as centers of innovation, worth to visit (the development of tourism and local business),
- The social aspect - to improve the quality of life with a possibility to faster movement; reducing the congestion problems that occur in urban areas.

The precursor of the idea of bike-sharing system was Amsterdam, where in 1965 Luud Schimmelpennink gathered hundreds of bicycles, painted them white and shared into the city, for free urban use. This project is known as the White Bicycle Plan. Spectacular success in the development of bike-sharing took place in 2007 in France. Then launched 750 automated rental of 10.000 bikes. In a short time, these numbers increased to 1.450 stations and 20.000 two-wheelers contributing to the onset of the global fashion for this type of urban transport. Now, there are over 140 systems bike rentals in 165 countries around the world. The biggest bike rental is situated in China's Hangzhou city, where residents could hire 60.000 bicycles [11]. The availability of bike sharing in the world was presented in Fig. 1.

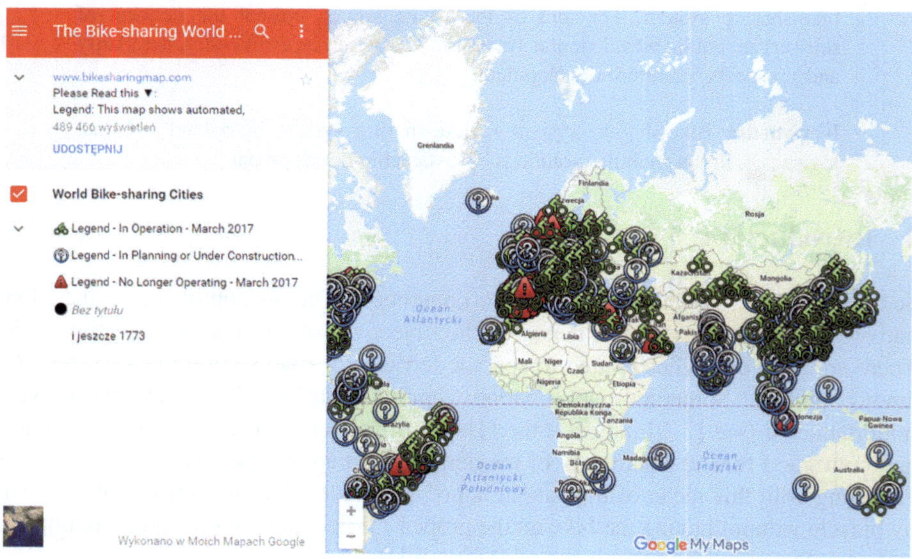

Fig. 1. The availability of bike-sharing systems in the world. Source: [12]

The origins of the occurrence of bike sharing in the Polish city dates back to 2008. Then Wavelo system became operational in Krakow. At the same time it took place an implementation of the bike-sharing systems in the following cities. Then the Polish inhabitants began to be mass-persuasion to use the concept and to replace standard travel by car to travel by bike. Introduced initially solutions were not, however, perfect. The cities lacked the right kind of system infrastructure, users complained about the shortage of docking stations where you can leave the bike, it was not possible to reserve the bike through mobile applications and verify in real time in which place on the map of the town the certain amount of bikes is available [13, 14]. Gradually, with the development of the concept and the appearance on the market of new, already experienced in foreign markets operators, the number of investments in urban public bicycle grew. Development of the system closely associated also with the emergence of newer and newer solutions, telematics and telecommunications [13]. With the development of IT solutions in the market appeared 4th generation of the bike-sharing system. This generation began to be defined as: 'bike mobility based on a fleet of bike sharing technical facilities and ICT (…)' [15].

What's more, cycling in Poland is growing steadily [16]. Bicycle culture raises new challenges in terms of traffic engineering as well as safety [17–19]. It should be noted, however, that bicycle rental systems make an important contribution to promoting this way of traveling in cities [20].

Currently, the system self-service bike rental includes more cities in Poland. The main operator of the system is mostly German company Nextbike. Selected cities with functioning system of bike-sharing and the biggest number of available bikes to share are presented in Table 1.

Table 1. Selected Polish cities with a functioning system of bike-sharing (Source: [21–26]).

City	System's name	Number of bikes	Number of dock-stations
Warsaw	Veturilo	3039	204
Kraków	Wavelo	1500	150
Łódź	Łódzki Rower Publiczny	1000	100
Poznań	Poznański Rower Miejski	923	88
Lublin-Świdnik	Lubelski Rower Miejski	900	90
Wrocław	Wrocławski Rower Miejski	760	76
Białystok	Białostocki Rower Miejski	450	45
Szczecin	Szczeciński Rower Miejski	338	33
Radom	Radomski Rower Miejski	250	25
Opole	Opole Bike	188	19
Bielsko-Biała	BBbike	120	12
Rzeszów	RoweRes	100	20
Katowice	City by Bike	88	11
Kalisz	Kaliski Rower Publiczny	24	3

3 Availability of Bike-Sharing Systems in Poland

The Season the availability of bike-sharing in most of Polish cities starts on 1st of March and ends usually on 30th of November. After this period of time, bikes are transferred to warehouses in order to preserve them. During that process, cities are equipped with other dock-stations to collecting bikes and prepared for the next cycling season. Prepared for the 2017 season bicycle dock-stations belonging to public bicycle system - City by Bike - in Katowice is presented in Fig. 2.

Fig. 2. The City by bike station, Katowice, Mariacka Street

Not all Polish cities cycling season is limited. An example of urban bicycle system, offering the possibility of using its services throughout the year is Krakow system solution called - Wavelo. The example of available bikes and racks in the system Wavelo of 27th of February 2017 is presented in Fig. 3.

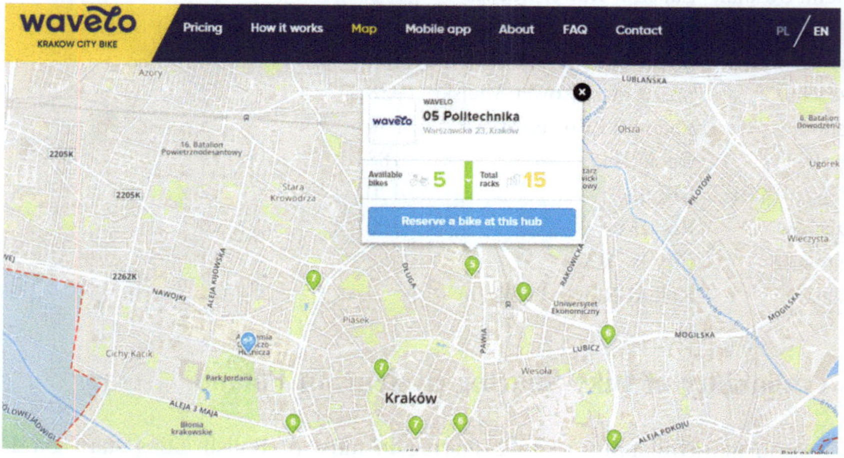

Fig. 3. Availability of bicycles and bicycle racks in Krakow's public bike system Wawelo of 27th of February 2017) Source: [27]

4 Popularity and Prospects of Development of Public Bicycle System in Poland

The popularity of self-service rental bikes is increasing from year to year [28]. It is no different in Poland. The biggest public bike rental system in Poland is Warsaw solution called Veturilo. The season 2016 was the fifth year of operation of the market. During the period of operation of the system from March 1 to November 30 of Warsaw residents benefit from them 1 866 423 times. The average time to rent a bike in the system was 25 min and 28 s.

Bicycles were rented for the longest period of time - 29 min in May, while the shortest in November - 16 min. Rent followed mainly in the afternoon i.e. between 16.00 and 19.00 [29].

An interesting fact is that the most active user of the Veturilo system borrowed bike 1025 times in season 2016 [21].

The Popularity of bike-sharing has not been decreased in Lodz, which is the second Polish bike rental-center. Here, in the cycling season from 30th of April 2016, when the Public Bike Lodz began operating, to 30th of November 2016 years users have made 1 501 881 of loans. Łódzki Rower Publiczny also broke the Polish daily record of bike loans which amounted to 17 665 [22].

Total lending of bike sharing was also high in the case of other cities with the implemented system. For the third, in terms of the quantities available for users of bicycles, the city - Lublin - the number of loans in 2016 amounted to 88 854. For other cities while Wroclaw - 1 008 184, Katowice - 38 217 [23–25].

The popularity of urban cycling is mainly evidenced by deploying it in other cities. The largest operator in Poland bicycle public company Nextbike with MultiSportBike program is currently available for residents of cities such as: Warsaw, Opole, Poznań,

Wrocław, Grodzisk Mazowiecki, Lublin, Konstancin Jeziorna, Białystok, Katowice,

Łódź, Stalowa Wola [26]. The popularity of bike-sharing in four Polish cities: Warsaw (Veturilo), Wrocław (Wrocławski Rower Miejski), Poznań (Poznański Rower Miejski), Opole (Opole Bike) with amount of loans from 2012 to 2016 is shown in Fig. 4.

In addition, its interest has attracted the rural population. The program now operates in the municipality of Choroszcz and municipalities Juchnowiec Kościelny [23].

According to the announcement by Nextbike, 2017 will be a momentous year for the public bicycle in Poland. Prospects for the development of the system are very high, especially since the program join another cities. Currently, the agreement to implement bike-sharing were signed with the cities: Sosnowiec, Gliwice and Radom [24].

Operators of public bike systems do not remain indifferent to the expectations from their users. An example would be the biggest system in Poland - Warsaw Veturilo that from 1st of March 2017 is expected to begin the new season in a completely different form. Development of the system is associated not only with its extension for another docking stations or optional equipment in a larger number of bicycles (including tandems, electric bikes and bicycles for children) but also a change from the point of view of the technological operation of the system. The new functionality of Veturilo maps, with a possibility to choose of the type of bike is presented in Fig. 5.

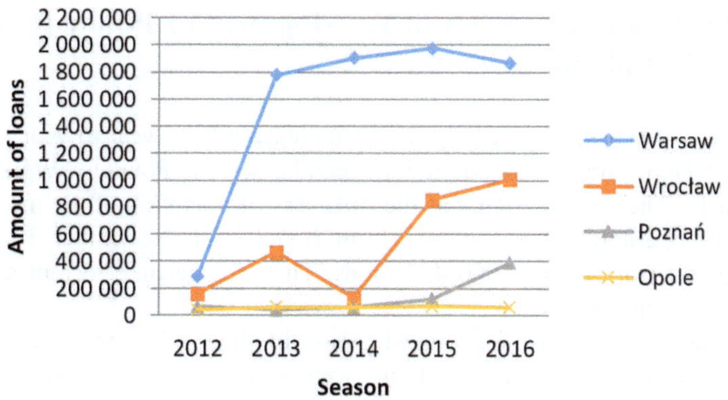

Fig. 4. The amount of bike loans in four Polish cities (Warsaw, Wrocław, Poznań, Opole) from 2012 to 2016 Source: [21, 36–39]

Fig. 5. The new functionality of Veturilo maps, with a possibility to choose of the type of bike-electric, standard, children, tandem Source: [30]

Also terminals and the methods of payment are upgraded. Terminals of the system have been adapted to make use of it by means of proximity cards. Docks will have LED, that will inform if the bicycle is ready to rent. Moreover, it becomes supplemented with universal pump cycle and the bicycles themselves are exchanged for new one. They are also more comfortable and equipped with saddle with a special gel coating that prevents soaking water into it. More comfortable driving position is provided by mounting a special ergonomic steering wheel with raised stands. Another form of amenities is also the use of a special graduation for seatpost mounting bicycle saddle, so that users can easily adjust the setting of the saddle to your posture and remember the number to use in the future. In the new season is going to be launched as a dedicated system mobile application that is allowed to rent a bicycle by scanning the QR code [30].

The creators of the system also care about the safety of their users. From the new season, cyclists will be required to provide your social security number PESEL or supplement it owned already used account [30].

5 Advantages and Disadvantages of Bike-Sharing System

Self-service bike rental system has many of his supporters and opponents. The basic advantages associated with bike-sharing can include economic benefits such as no need to purchase a bicycle on the property or service required it.

Another of the advantages of the bicycle system is no need to have their own parking space. Another benefit is the opportunity to travel by bike only during the one way that is called 'asymmetric travel' [15].

Using solutions of bicycle sharing system also brings benefits to the city, in which the system is valid. The use of such solutions is an opportunity to contribute to the perception of the center as an innovative, or even characterized by elements characteristic of the smart city.

This solution has a chance to increase a value for tourists, especially those who have positive attitude to any ecological and pro-social solutions. It could also bring benefits in the form of promotion of the city as well as the development of local businesses [29, 31–33].

One of the drawbacks depending on the system is not adjusting the offer to users of all ages. In most systems, there is no possibility to loan a bike adapted to traveling for children. In addition, often the defect is also insufficient infrastructure of the system, which means the shortage of docking stations.

Systems operators of lower rank popularity often yet functioning on the basis of operational models of previous generations, i.e. without the use of advanced ICT. Then checking the bike availability or information on the website and the state of the actual availability of the bicycle may be different [31, 34, 35].

The disadvantage of the system is the lack of customization bicycle options or absence the required accessories such as e.g. a bike child seat or bike basket.

6 Summary

As is shown by the presented statistics, the popularity of public bicycle lending in the Polish cities begins to be measured in millions. Using the services of bike-sharing allowed us to bridge the gap between the supply of standard transport. In addition, the implementation by the city bike system is for local governments much more economical alternative than building new tram network or planning another bus line connections. In summary, the bike-sharing services have a chance to become the ideal alternative means of transport, especially the transport of the first and last mile. Solutions of bike-sharing systems may become the perfect complement to the logistics of Polish cities and a way to promote alternative forms of urban mobility.

References

1. Banister, D.: European Transport Policy and Sustainable Mobility. Taylor & Francis, London (2000)
2. European Commission: Accompanying the White Paper – Roadmap to a Single European Transport Area – Towards a Competitive and Resource Efficient Transport System. 144 Final. European Commission, Brussels (2011)
3. European Commission: Communication from the Commission to the European Parliament, the Council, the European Economic and Social Committee and the Committee of the Regions. Clean Power for Transport: A European Alternative Fuels Strategy. European Commission, Brussels (2013)
4. Szczuraszek, T., Macioszek, E.: Proportion of vehicles moving freely depending on traffic volume and proportion of trucks and buses. Baltic J. Road Bridge Eng. **8**(2), 133–141 (2013)
5. Macioszek, E.: Relationship between vehicle stream in the circular roadway of a one-lane roundabout and traffic volume on the roundabout at peak hour. In: Mikulski, J. (ed.) Telematics - Support for Transport. CCIS, vol. 471, pp. 110–119. Springer, Heidelberg (2014)
6. Macioszek, E.: Analysis of the effect of congestion in the lanes at the inlet to the two-lane roundabout on traffic capacity of the inlet. In: Mikulski, J. (ed.) Activities of Transport Telematics. CCIS, vol. 395, pp. 97–104. Springer, Heidelberg (2013)
7. Staniek, M.: Road pavement condition as a determinant of travelling comfort. In: Sierpiński, G. (ed.) Intelligent Transport Systems and Travel Behaviour. AISC, vol. 505, pp. 99–107. Springer, Switzerland (2017)
8. Sierpiński, G., Staniek, M., Celiński, I.: New methods for pro-ecological travel behavior learning. In: 8th International Conference of Education, Research and Innovation ICERI Proceedings, pp. 6926–6933, Seville (2015)
9. Staniek, M.: Stereo vision method application to road inspection. Baltic J. Road Bridge Eng. **12**(1), 38–47 (2017)
10. Bike-Share Opportunities in New York City. https://a841-tfpweb.nyc.gov/bikeshare/files/2014/08/bike_share_complete.pdf
11. Wrower Service. The Urban Bike-Rental Companies. https://wrower.pl/miasto/rower-miejski-wypozyczalnia-rowerow,2183.html
12. Google Maps. The Bike Sharing Systems in the World. https://www.google.com/maps/d/u/0/viewer?mid=1UxYw9YrwT_R3SGsktJU3D-2GpMU&hl=en
13. Midgley, P.: Bicycle-sharing schemes: Enhancing sustainable mobility in urban areas. https://www.cleanairinstitute.org/cops/bd/file/tnm/13-bicycle-sharing.pdf
14. Website. https://www.transport-publiczny.pl/wiadomosci/rower-miejski-w-polskich-miastach-jak-dziala-1730.html
15. Taniguchi, E., Tompson, R.G.: Innovations in City Logistics. Nova Science Publisher, New York (2008)
16. Okraszewska, R., Birr, K., Gumińska, L., Michalski, L.: Growing role of walking and cycling and the associated risks. Proceedings of Gambit. Gdańsk (2017)
17. Nosal, K.: Travel demand management in the context of promoting bike trips. An overview of solutions implemented in Cracow. Transp. Probl. **10**(2), 23–24 (2015)
18. Okraszewska, R., Jamroz, K., Bauer, M., Birr, K., Gobis, A.: Identification of risk factors for collisions involving cyclists based on Gdańsk example. Environmental Engineering Proceedings (2017)

19. Okraszewska, R.: The impact of organizational culture on bicycle commuting frequency. the research based on example of three IT companies. In: Bąk, M. (ed.) Transport Development Challenges in the Twenty-first Century. SPBE, pp. 201–210. Springer, Cham (2016)
20. Okraszewska, R., Jagielska, Z.: Comparison of Polish and Danish Approach to Promotion of Cycling. Environmental Engineering Proceedings (2017)
21. Veturilo Bike-Sharing System. The Five Years of Bike Revolution in Warsaw. https://www.veturilo.waw.pl/2016/12/07/piec-lat-rowerowej-rewolucji-w-warszawie-za-nami-sezon-2016-rowerow-miejskich-veturilo
22. Lodz Bike Sharing System. The Urban Bikes Revolutionize Lodz city. https://lodzkirowerpubliczny.pl/rowery-publiczne-zrewolucjonizowaly-lodz-15-miliona-wypozyczen-pod-koniec-pierwszego-sezonu
23. Lublin Bike Sharing System. Last Day of the Bike Sharing Season. https://lubelskirower.pl/news/dzis-ostatni-dzien-sezonu-lubelskiego-roweru-miejskiego-rekordowe-850-tysiecy-wypozyczen
24. Wroclaw Bike Sharing System. The Best Bike Sharing Season Comes to end. https://wroclawskirower.pl/news/dobiega-konca-rekordowy-sezon-wroclawskiego-roweru-miejskiego-peklo-milion-wypozyczen-w-ciagu-jednego-roku
25. Citybybike System. The summary of the best urban bike season. https://citybybike.pl/news/oficjalne-podsumowanie-najlepszego-sezonu-w-historii-rowerow-miejskich-city-by-bike-170-dni-65-tysiaca-nowych-uzytkownikow-i-blisko-40-tysiecy-wypozyczen
26. The Nextbike Bike-Sharing Operator. https://nextbike.pl/multisportbike
27. Wavelo Bike-Sharing System. https://en.wavelo.pl#map
28. The Gdańsk Metropolis. The concept of bike sharing system. https://www.metropoliagdansk.pl/upload/files/Studium_koncepcyjne_SRMOMGGS%20wersja%204_0.pdf
29. Ricci, M.: Bike Sharing: A review of evidence on impacts and processes of implementation and operation. Res. Transp. Bus. Manag. **15**, 28–38 (2015)
30. The Nextbike Bike-Sharing Operator News. https://nextbike.pl/news/
31. Cohen, B., Kierzmann, J.: Ride on! mobility business models for the sharing economy. Organ. Environ. **27**(3), 279–296 (2014)
32. Parkin, J., Wardman, M., Page, M.: Estimation of the determinants of bicycle mode share for the journey to work using census data. Transportation **35**(1), 93–109 (2008)
33. Miciukiewicz, K., Vigar, G.: Mobility and social cohesion in the splintered city: challenging technocentric transport research and policy-making practices. Urban Studies **49**, 1941–1957 (2012)
34. Murphy, E., Usher, J.: The role of bicycle-sharing in the city: analysis of the irish experience. Int. J. Sustain. Transp. **9**, 116–125 (2015)
35. Okraszewska, R., Grzelec, K., Jamroz, K.: Developing a cycling subsystem as part of sustainable mobility strategy – the case of Gdansk. Sci. J. Silesian Univ. Technol. Series Transp. **90**, 11–21 (2016)
36. The Veturilo System, Summary of the 2012-2013 Season. https://www.um.warszawa.pl/sites/default/files/attach/aktualnosci/veturilo_-_podsumowanie_sezonu_final.pdf
37. The Opole Bike System. https://nextbike.pl/miasta/opole-bike/o-opole-bike/
38. The Wroclaw City Bike System. https://wroclawskirower.pl/
39. The Poznań City Bike System. https://nextbike.pl/miasta/poznanski-rower-miejski

Privatization of Entities in the Air Transport Sector - Airports

Stanisław Miecznikowski and Dariusz Tłoczyński(✉)

Faculty of Economics, University of Gdańsk, Gdańsk, Poland
stanmiecznik@gmail.com, dariusz.tloczynski@ug.edu.pl

Abstract. The privatization process means a change in ownership proportion in the economy for the benefit of the private sector. The paper develops the theoretical models to examine the relationship between ownership changes in the air transport industry in the selected world airports and Poland. We tried to study the cases of the recent privatization of the both airports The results are consistent to get a right for buying assets with the view of the expanding operations and the benefiting from the network effects offered by privatization.

Keywords: Privatization · Airport · Air transport

1 Introduction

The issue of ownership can be referred to the institutional and legal forms of enterprises. Apart from companies that are formally owned by individual persons, commercial law companies can be distinguished: personal and capital ones [1]. On the Polish air transport market, the dominant form of doing business are capital companies: limited liability companies and joint stock companies. However, it should be pointed out that there are state-owned enterprises operating under the law on state-owned enterprises [2].

The main objective is to indicate the directions of privatization of airports. Based on the theoretical and practical assumptions, the most beneficial methods of ownership transformation were chosen.

2 Why Privatization?

The privatization process means a change in ownership proportion in the economy for the benefit of the private sector. This is usually the combined effect of two separate processes. The first is the transformation of existing public property (state and municipal) into private ownership. The second is the emergence and development of private entities, both domestic and with foreign capital.

With regard to airports, the privatization process involves the temporary delivery to the private sector of the use of aviation infrastructure in order to raise funds to finance investments related to increase the capacity of infrastructure [3].

The theoretical arguments for and against privatization of publicly owned organizations, particulary when a share flotation is being considered, are well known.

E. Macioszek and G. Sierpiński (eds.), *Recent Advances in Traffic Engineering for Transport Networks and Systems*, Lecture Notes in Networks and Systems 21,
DOI 10.1007/978-3-319-64084-6_16

They have been fiercely debat over the years and are well documented [4, 5]. Privatization will reduce the need for public sector investment and provide access to the commercial markets. It will reduce government control and interference and may increase an organization's ability to diversify. It may bring about improved efficiency, greater competition, wider share ownership and provide greater incentives for management and employees to perform well. On the other hand, it may create a private monopoly which overcharges, delivers poor standards of service or invests inadequately [6].

The nature of airport privatization:

- theoretical analyses and empirical evidence suggest that private ownership is most efficient,
- simplistic view of privatization: change of ownership of the property and facilities and specifically the transfer from a government agency to private exception,
- misleading regarding airports, since most privatizations didn't involve the actual sale of the property - the BAA being the most notable exception,
- airports privatization usually involves only the transfer of some ownership rights [7].

Literature sets out the main objectives of privatization of entities in the air transport sector. These include:

- efficient and economical aims:

- greater operational and financial efficiency,
- the ability to raise capital from external sources more quickly for investment in infrastructure,
- relieving the state budget of expenditures on the development of airports,
- generating income on the state budget from the sale or lease of airport assets,
- the risk transfer (among others decreasing the risk of investing without reasonable economic efficiency),

- social:

- customer orientation - providing better service standards,
- keeping existing workplaces and creating new ones,
- improving the quality of service for passengers,

- ideological:

- political popularity,
- limiting the role of the state in economic life for the benefit of private institutions,

- strategic:

- constructing a system that will ensure the development of airports to a degree that will enable the region's continued economic development [8–10].

In systemic transformations aimed at the privatization of operators of airports R.W. Poole distinguishes two strategies:

- privatization of the company as a whole, while creating legal and institutional conditions for regulating the sector (privatization without restructuring),
- division of an integrated state-owned enterprise in order to create a few companies capable of independent functioning on the market and mutual competition (first restructuring and setting of market rules, then privatization [11].

3 The Nature of Privatization Airport

The economic deregulation of airport has produced important challenges. With increasing competition charges have dropped and demand has increases considerably putting significant pressure on existing airport infrastructure.

Consequently, air traffic congestion and delays have the world wide consensus. The authorities in chargé of sector have considered not only regulatory reform of airport, but also a charge of ownership as possible solutions to the problem of airport congestion and in some case, a means of facilitating expansion of airport capacity. Airport privatization has become a world wide phenomea since the classic sale of the British Airport Authority (UK) in 1987 [12]. Since them 1990 many emerging countries have entered into short, and long term airport privatization transaction. Examples includes airport in Delhi, Mumbai and Bangalore in India, and Beijing and Shanghai in China.

Airports nowadays are generally run as modern business rather than public utilities [13]. This new trend in the global airport industry means that along with delivering airside services to airlines, airports have to considerable lengths to among other things, attract new services, and maintain high service level and low operating cost to enable them to face competition from other airports and transports models as well as to maximize the generation of nonaeronautical revenues from terminal retail services, increase acceptability and transparency to investors and develop vertical relations with airlines and global airline alliances.

In this context, airport privatization has been regarded in many cases as a tool not only for security proper financing but also for introducing improvements in airport management.

There is an analytical stream of the literature that studies airports. Some recent literature investigated airport privatization from a modeling perspectives [14–19], among others.

As such, airports must add value to their main client groups, airlines and passengers. A two side platform is capable of exploiting the demand interdependencies of it's clients and allows direct and indirect network effect to emerge or to be more intensively generated. Airlines benefit if a two sided platform airport increases the number of passenger benefit if the airport attract more airlines, more destinations and higher flights frequencies.

We investigated whether of the management and as an outcome, enhances the dynamic capabilities of airports, so their more closely resemble a two sided platform [20]. Employed the concept of dynamic capabilities to describe, firms with timely responsiveness and rapid flexible product innovation. We suspect that privatization may mark the birth of an entity that acts as facilitator allowing for previously unexploited direct and indirect network effects between airlines and passengers to energy.

We test the hypotheses related to the consolidation of privatized airports as two side platforms try empirically examining the possible impact of privatization airline demand. We examine whether privatizations allows for market demand shift as associated with the emergence of network effects and the possible positive feedback loops triggered by them.

To date, the existing empirical literature applied to airports is scare. Invalid et al. 2012, who weather first to develop an empirically framework to investigate the potential role of airports as the two side platforms. Bettini and Olivera, [21] raised the hypothesis that carriers may anticipate the benefits and the entry inducement effects associated with the enhanced two side platform on the case of recent privatization of Brazilian airports and control for the sequence of public events with the airport privatization.

4 Polish Airport Industry in 2015

Until 1987, airports in Poland operated under the centralized Air Traffic and Airports Management. As a result of market changes at the beginning of the 1990s, privatization of Polish airports took place on the basis of the law regulating ownership relations, thus the second strategy was fulfilled [22].

As a result of market changes, government institutions became the owners/operators of airport operators - government/municipal privatizations and public and private companies [23]. The legal form of airports also changed.

In the European Union countries, most airports are dominated by public ownership - state or local authorities. In Poland, all airports are public companies within the meaning of the Article 106 of the Act 1 Treaty on the Functioning of the European Union. In all airports, except for Szymany, public ownership dominates the structure of capital. The structure of shares in Polish airports is shown in Table 1.

At most airports one of the shareholders is the “Polish Airports” State Enterprise, but the involvement of local authorities in the development of the local transport market is increasing systematically. There is a strong relationship between regional and local development of settlement structures and the development of airports. Cities are interested in the development of the aviation services market and often local authorities support aviation infrastructure development initiatives in the region. An example is included the opening of new regional airports in Modlin and Lublin in 2012.

Table 1. The structure of legal and management issues Polish airport industry for 2015 (Source: [24–36])

Airport	Legal form	Share capital in PLN	Owners in %			
			City	Region	Enterprise	Polish enterprise airports
Bydgoszcz BZG	S.A.	73549710	22.62 - Bydgoszcz 0.05 - Toruń 0.01 - Inowrocław ~0.00 - Sicienko	68.50	1.20	7.62
Gdańsk GDN	Sp. z o. o.	125290000	33.62 - Gdańsk 2.23 - Gdynia 2.19 - Sopot	32.89	0	29.09
Katowice KTW	S.A.	137099300	4.89 - Katowice	34.87	42.48	17.30
Kraków KRK	Sp. z o. o.	101232000	1.04 - Kraków 0.04 - Zabierzów	22.73	0	76.19
Lublin LUZ			58.42 - Lublin 5.39 - Świdnik	36.19	0	0
Łódź LCJ	Sp. z o. o.	162737500	94.29 - Łódź	5.71	~0.00	0
Modlin WMI	Sp. z o.o.	323824500	4.91 - Nowy Dwór Mazowiecki	30.37	34.43	30.39
Poznań POZ	Sp. z o. o.	290385000	37.00 - Szczecin	24.00	0	39.00
Rzeszów RZE	Sp. z o. o.	470642600	0	53.78	0	46.22
Szczecin SZZ	Sp. z o. o.	135550000	33.39 - Szczecin 3.35 - Goleniów	13.52	0	49.74
Szczytno SZY	Sp. z o. o.	10633333	3.57	5.49	74.20	12.67
Warszawa WAW	PP		0	0	0	100
Wrocław WRO	S.A.	206830000	31.11	0	19.74	31.11
Zielona Góra IEG	Sp. z o. o.	380000	0	100	0	0

5 Types of Privatization

ICAO is following forms of ownership and models of airport management:

- government ownership,
- management contract,
- lease or concession,
- transfer of minority ownership,

- public-private partnerships, and
- private sector ownership and control [37].

Types and examples of airport privatizations all over the world are presented in Table 2.

Table 2. Types and examples of airport privatizations all over the world (Source: [40])

Charateristic		Criteria:	Examples
Share flotation	Government/municipality sells airport shares fully or partially through an IPO (Initial Public Offering) - the shares are trade dat one or more stock exchanges	- Government / municipality gives up total partial ownership, - Private sector (institutional investors and the public)own airport shares, - Airports company management is responsibility for operations, profitability and investment, - Shareholders take the full economic risks.	BAA - 100% IPO, VIA - 27% IPO, CPH - 25% IPO, BCIA (China) - 35% IPO, Auckland (New Zealand) - 52% IPO, FRA - 29% - 29% IPO.
Trade sale	Government / municipality sells airport shares fully or partially to a strategic partner with the option of immediate or later share flotation	- Strategic interest prevails over financial interests, - Major incentive is the known - how transfer from the strategic partner, - Strategic partner has operational dominance through management contract, - Strategic partner shares economic risk with the remaining shareholders.	Sanford, Orlando, Wellington, (New Zealand), South Africa, Humberside (UK), Hamburg (Germany).
Concession	Purchase of a concession to run the airport for a defined period	- Upfront payment, - Option for renewal, - Responsibility for operations and future investment, - Concessionare takes full economic risk.	AENA, YVRAS, M.Macdonals, Fraport, BAA, VIE.
Project finance privatization	Franchise to build and operate the airport or parts of it for a defined period and than transfer the facility back to owner	- No upfront payment, - Operator retains all revenues, bears all cost and builds the facility, - Operator takes full economic risk for investment and operations.	Athens (Greece) - Greek State/Hochtief, New York JFK (USA) - Schiphol consortium, Sharm El Sheikh (Egypt) - YVRAS consortium.
Management contract	Government / municipality grants a management contract to operate the airport or parts of it for a defined period of time	- Ownership remains with Government / municipality, - Remuneration based on fixed as well as result oriented fee, - Manager splits economic risk with owners, - Investments are made by the owners, - Option fot renewal.	BAA - Pittsburgh, Boston, Indianapolis, AENA - Cuba and Columbia, AGI/TBI - Burbeck, Atlanta, AerRianta - duty-free at airports in Russia and Middle East, YVRAS - Bermuda, Cook Islands, Turks and Caicos Islands, ADP - Cameroon and Madagascar.

After 25 years of operation of airlines with majority share of public capital, there is a need for further privatization and restructuring of Polish air transport companies. Krajewski [38] believes that further privatization will enable and support restructuring processes in enterprises, which in turn may prove to be a means of increasing the competitiveness of enterprises in a competitive market. Graham presents five models of privatization:

- share flotation,
- trade sale,
- concession,
- project finance privatization,
- management contract [39].

A similar classification is given by Poole [11, 41]. In turn, Kapur [42] makes more detailed divisions. It distinguishes privatization techniques such as managerial contract, B-O-T, B-O-O-T with variants, L-D-O and long-term leasing, redemption and entry into the capital market. The authors analyzes the privatization of Polish airports based on the models presented by Vogel and Graham. They are most often quoted in the literature related to privatization in the aviation sector.

6 Some of Remarks About the Privatization Polish Airports

Polish airports function as public utility airports. Only the biggest one are the most profitable. Thus, three concepts of privatization should be pointed out. The first concern should be the "Polish Airports" State Enterprise in Warsaw, the second large regional airports, mostly with positive financial results and third small regional airports.

The Ministry of Infrastructure and Development is desired commercialize the "Polish Airports" State Enterprise. The project assumes transformation into a joint stock company, although the only shareholder would be the State Treasury. The author of the paper does not share the position of the Ministry of Infrastructure and Development. From a business model point of view, the "Polish Airports" State Enterprise's continued existence is pointless. Such a view is shared by a part of the academic community as well as experts in the field of air transport. It is proposed that the shares of the "Polish Airports" State Enterprise in regional ports should be sold or transferred to regional authorities. Warsaw Chopin Airport should be sold to an entity specializing in airport management. What's more, the airport in Zielona Góra should be handed over to the marshal of Lubuskie Voivodship. This solution will provide the Warsaw airport with capital, access to know-how, development of the network of foreign connections and, above all, investment opportunities.

With regard to airports in Crakow, Gdańsk, Katowice, Wrocław and Poznań, the concept of privatization should concern increasing the effectiveness of organization management and the introduction of new instruments contributing to the quality of the services offered. The best solution for this group of airports would be a 15–20 years concession or managerial contract. Such solutions were successful in foreign airports. This mechanism causes the public entity to retain the ownership of the airport while entrusting operational and management activities to the private sector.

For small regional airports, a concession method, a management contract, or a BOT formula would be preferable. Such solutions contribute to the efficient management of the potential air transport market, but above all they transfer the responsibility for the financial result to the investor while maintaining the airport property rights. Nevertheless, in Polish conditions the proposed variants of privatization of this group of airports, may lack interest of potential investors. In this situation, the only solution could be selling to business.

7 Summary

It should be noted, however, that the opinions on the real effects of the privatization of Polish airports will be various. Pool Jr., Vogel, Dennis believe that privatization contributes to increased capital efficiency, while the Air Transport Research Society states that there are no significant differences between privatized and state-owned airports [43]. However, the involvement of capital forces investors to pursue activities aimed at achieving a high competitive position. This situation can be seen in many foreign airports.

In conclusion, the privatization of regional airports is dependent on local government. It should be noted that the privatization of air transport sector involves risks on the one hand for investors, on the other hand for public entities owning or co-owning airlines. Undoubtedly, the privatization process will result in the development of air transport companies, which will lead to the development of competitive processes on the industry market.

References

1. Law Office of the Parliament: Commercial Law, Act of 15 September 2000, Dz.U. 2013, pos. 1030 with Later Changes. Law Office of the Parliament, Warsaw (2000)
2. Law Office of the Parliament: Act of 23 October 1987 - Polish Enterprise Airport. Dz.U. 1987, pos. 185. Law Office of the Parliament, Warsaw (1987)
3. Wells, A.T., Young, S.: Airport Planning and Management. McGraw-Hill, New York (2004)
4. Jackson, P.M., Price, C.P.: Privatization and Regulation. Longman, London (2007)
5. Beesley, M.E.: Privatization, Regulation and Deregulation. Routledge, Abingdon (2013)
6. Graham, A.: Managing Airports. An International Perspective. Elsevier, Oxford (2008)
7. Vogel, H.A.: Impact of privatisation on the financial and economic performance of European airports. Aeronaut. J. **110**, 197–213 (2006)
8. Augustyniak, W.: Privatization of airports as a way for air infrastructure development. J. Int. Stud. **3**, 38–39 (2010)
9. Vasigh, B., Fleming, K., Tacker, T.: Introduction to Air Transport Economics, from Theory to Applications, 2nd edn. Routledge, Abingdon (2008)
10. Pitt, M.: Strategic direction in the airport business: enabling or disabling? Facilities **19**, 15–156 (2001)
11. Airport Finance: https://trid.trb.org/view.aspx?id=367503/
12. Graham, A.: The objectives and outcomes of airport privatisation. Res. Transp. Bus. Manag. **1**, 3–14 (2011)
13. Gillen, D.: The evolution of airport ownership and governance. J. Air Transp. Manag. **17**, 3–13 (2011)

14. Mantin, B.: Airport complementarity: private vs. government ownership and welfare gravitation. Transp. Res. Part B **46**, 381–388 (2012)
15. Assaf, A.G., Gillen, D.: Measuring the joint impact of governance form and economic regulation on airport efficiency. Eur. J. Oper. Res. **220**, 187–198 (2012)
16. Lin, M.H.: airport privatization in congested hub—spoke networks. Transp. Res. Part B **54**, 51–67 (2013)
17. Czerny, A.I.: public versus private airport behavior when concession revenues exist. Econ. Transp. **2**, 38–46 (2013)
18. Gillen, D., Mantin, B.: The importance of concession revenues in the privatization of airports. Transp. Res. Part E **68**, 164–177 (2014)
19. Yan, J., Winston, C.: Can private airport competition improve runway pricing? The case of San Francisco Bay Area Airports. J. Public Econ. **115**, 146–157 (2014)
20. Teece, D., Pisano, G.: The dynamic capabilities of firms: an introduction. Ind. Corp. Change **3**, 537–556 (1994)
21. Bettini, F.A.J., Oliveira, H.V.M.: Two-sided platforms in airport privatization. Transp. Res. Part E **93**, 262–278 (2016)
22. Law Office of the Parliament: Act of 13 June 1990—Privatization Polish Enterprises, Dz.U. 1990, pos. 298. Law Office of the Parliament, Warsaw (1990)
23. Krajewski, S.: The Economic Effects of Direct Privatization on the Basis of Research of 200 Polish Enterprises. In: Materials from the Conference Organized by the Legislative Commission and the National Economy Committee. The role of Direct Privatization in the Process of Ownership Transformation of the Polish Economy. Sejm of the Republic of Poland, Warsaw (1999)
24. Chopin Airport: http://www.lotnisko-chopina.pl
25. Krakow Airport: http://www.krakowairport.pl
26. Katowice Airport: http://www.katowice-airport.com.pl
27. Gdansk Airport: http://www.airport.gdansk.pl
28. Wrocław Airport: http://www.airport.wroclaw.pl
29. Poznan Airport: http://www.airport-poznan.com.pl
30. Lodz Airport: http://www.airportlodz.pl
31. Bydgoszcz Airport: http://www.plb.pl
32. Szczecin Airport: http://www.airport.com.pl
33. Rzeszow Airport: http://www.lotnisko-rzeszow.pl
34. Zielona Gora Airport: http://www.lotnisko.lubuskie.pl
35. Modlin Airport: http://www.modlinairport.pl
36. Lublin Airport: http://www.portlotniczy.lublin.pl
37. Commercialization and Privatization of Airports and Air Navigation Services Providers: http://www.icao.int/Meetings/atconf6/Documents/WorkingPapers/ATConf6-wp006_en.pdf
38. International Civil Aviation Organization: Privatization in the Provision of Airports and Air Navigation Services. International Civil Aviation Organization, Montreal (2002)
39. Krajewski, S.: Privatization, Restructuring. Competitiveness of Polish Enterprises. Polish Economic Publishing House, Warsaw (2009)
40. Vogel, H.A.: Airport Privatization: Is the Model Broken?. University of Westminster, Marylebone (2001)
41. Graham, A.: Managing Airports. An International Perspective. Elsevier, Oxford (2008)
42. Kapur, A.: Airport Infrastructure. The Emerging Role of the Private Sector. The International Bank for Reconstruction and Development, Washington (1995)
43. University of British Columbia: Centre for Transportation Studies, Air Transport Research Society: Global Standards for Airport Excellence. Centre for Transportation Studies, Columbia (2002)

The Problems of Optimisation of Transport Networks

The Impact of the Analysis Period on the Estimation Accuracy of Queue Lengths at Intersection Inlets with Traffic Lights

Tomasz Szczuraszek(✉) and Damian Iwanowicz

University of Technology and Life Sciences, Bydgoszcz, Poland
{tomasz.szczuraszek,damian.iwanowicz}@utp.edu.pl

Abstract. In this article, the authors described the problem of determining the traffic conditions at intersections with traffic lights in the state of traffic oversaturation. Based on their own research results gathered at three intersections in different traffic density states, an accuracy analysis was conducted for vehicle queue lengths calculation at the inlets of those intersections during various assumed sampling periods. The authors also conducted an accuracy evaluation of the estimation of vehicle queue lengths by means of several theoretical methods used in the USA, Germany, Great Britain, Australia, Canada, and Poland. That evaluation shows that in the case of traffic oversaturation states that last longer, errors in estimation using those methods are significant, regardless of the assumed sampling length. Based on the results of analyzing the obtained errors, the authors suggest assuming the maximal sampling time of a 15-min interval, while they recommend a 5-min one.

Keywords: Queue lengths · Urban signalized intersections

1 Introduction

The basic means of transport for dwellers of metropolises, as well as large and medium-sized cities, in economically developed countries, is an automobile. Thanks to its numerous advantages (above all travel comfort, time, and directness), it has been dominating for over a quarter of a century when it comes to travelling. Due to the low cost of second-hand vehicle purchase, it is also easily available in all of the above mentioned countries. In the modal division of means of transport used in passenger travel in the EU countries, an automobile constitutes 71% [1], and in the USA and Poland around 80% [2, 3]. As a consequence, problems connected with high crowding of the basic traffic networks in the key periods of carrier peaks occur in the majority of the analyzed cities. The consequence of the overcrowding of urban traffic networks is in turn the constitution of permanently long vehicle queues, especially at the inlets of intersections with traffic lights. Because of that, individuals moving within a traffic network are subject to more and more significant travel time losses, as well as a bigger stress load, while also contributing to higher fuel usage, increasing the environmental pollution and traffic noise levels.

E. Macioszek and G. Sierpiński (eds.), *Recent Advances in Traffic Engineering for Transport Networks and Systems*, Lecture Notes in Networks and Systems 21,
DOI 10.1007/978-3-319-64084-6_17

The observed conditioning of the escalation in traffic crowding in cities allows to form an assumption that it is highly probable that in the near future, one will still be forced to accept the significant travel time losses, both nationally and worldwide. The main reason for that are certainly traffic-incapable road networks, mostly including intersections with traffic lights, which almost always pose as "bottleneck" in a basic network system. It needs to be stressed that the deteriorating state in traffic conditions is significantly increasing beyond the traffic peaks, which has been observed for a longer time.

Undoubtedly, in order to search for effective ways to improve the traffic conditions in the crowded road networks, one needs to know the process of the traffic flow in such networks more thoroughly, especially the process of forming and crowding of vehicle queues in front of intersection inlets with traffic light sets. It needs to be remembered (which a lot of designers and planners tend to forget) that the degree of traffic load (volume-to-capacity ratio) at intersection inlets in the state of traffic oversaturation can only be properly estimated by including the length of queues formed at the beginning of the red light signal on individual lane groups:

$$X = \frac{Q + \frac{k_O}{t_a}}{C} \tag{1}$$

where:

X - volume-to-capacity ratio of a traffic lane, a traffic lanes calculation group, an inlet, an intersection, respectively [-],

Q - demand flow rate of vehicles arriving at the traffic lane, a traffic lanes cal culation group, an inlet, an intersection, respectively in [veh./h],

C - capacity of traffic lane, a traffic lanes calculation group, an inlet, an inter-section, respectively in [veh./h],

k_O - initial queue length, which occurs before the analyzed time period [veh.],

t_a - duration of the analysis period in [h].

Incorrectly determined vehicle queue lengths, both initial (remaining from previous traffic cycles) "k_O" and maximum ones "K_M" (occurring at the beginning of the green light display) cause not only false estimation of the volume-to-capacity ratio at the inlet, but also errors in determining other traffic conditions indicators, such as time loss, number of stops, exhaust gas environmental pollution level, etc. It also has a major influence on the correct match of the traffic signal control systems, especially including the ITS, as well as determining the number and length of traffic lanes.

One of the significant problems with estimating the vehicle queue lengths k_O and K_M is the change in traffic conditions over time in the individual sections of a crowded road network, concerning mainly the volumes of traffic outflow and traffic inflow. This depends on both deterministic and random factors. In the case of small volume-to-capacity states, such fluctuation does not affect the forming of remaining vehicle queues significantly. However, such queues will occur after a certain boarder-line volume-to-capacity state is exceeded. Should such a state last longer, vehicles remaining in front of an intersection inlet during individual traffic light cycles will accumulate and, ipso facto, both k_O and K_M will increase. This process shows that small errors in the estimation of remaining vehicle queues after every individual traffic light cycle may lead to significant errors in the estimation of the final value of such queues, which constitute the sum of the vehicles that were not served in the individual cycles from the vehicles which arrived during those cycles.

In view of the above mentioned facts both the estimation accuracy of the queue remaining from a single traffic light cycle and the assumed analysis period are of great importance. It is obvious that prolonging of the analysis time beyond one traffic lights cycle will always cause averaging of the estimated values and it will be the greater, the longer the assumed analysis period. The fluctuations of the vehicle queue lengths values occurring in individual cycles will then be omitted. Undoubtedly, this influences the queue lengths estimation errors. In some cases, overly averaging the values of remaining queue lengths may make it impossible to define the moment of when a traffic jam is created. An additional difficulty in the estimation of queue lengths is the fact that the increase of a queue at the inlet occurs faster than its discharge. This is due to the fact that the volume of vehicles stopping in front of every inlet is higher than the volume of vehicles moving away from the queue at these inlets (Fig. 1).

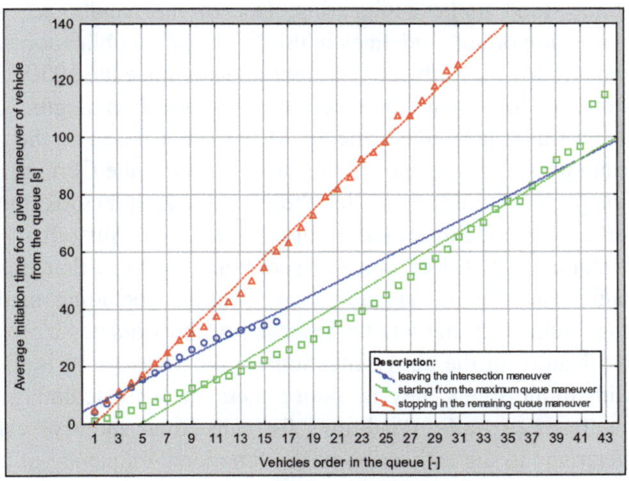

Fig. 1. An example comparison of average time intervals between vehicles in the process of vehicles leaving an intersection inlet, vehicles starting from the maximum queue and vehicles stopping in the remaining queue

In this dissertation, the authors attempt to determine the impact of the analysis period on the estimation errors of queue lengths at intersection inlets with traffic signal control systems, based on their own research.

The research on the vehicle queue lengths was conducted in 2016 in the city of Bydgoszcz (a city of the size of approx. 176 km^2 and the population of approx. 355 thousand), at three intersection inlets with traffic lights, placed in a basic city road network system. The research was conducted using 10 digital cameras, installed together with power supply equipment along an intersection inlet and the area leading to that inlet. The recorded video was then merged and synchronized, so that reading and analyzing the data could be conducted using only one large screen.

The objective of the research was to specify the empiric lengths of vehicle queues forming during single consecutive traffic signal cycles of:

- queues forming during the red light signal (at the end of the red interval) K_C,
- queues remaining after the green light signal (residual queue lengths) K_P,
- maximum back-of-queue size falling on a given traffic signal cycle (maximum queue lengths) K_M.

Altogether, almost 2 thousand traffic signal cycles on 5 lanes were analyzed.

2 The Currently Recommended Analysis Period Lengths in the Estimation of Traffic Conditions at Signalized Intersections

Any analyses of traffic conditions are most often conducted based on project guidelines from broadly understood traffic engineering. One of the handbooks, considered by many to be the most important guidelines in this field, is the publication called the High Capacity Manual (HCM) [4] - subject to current updates since the 1950s. Depending on the local conditioning of a particular country (motorization degree, road network development, local traffic laws, etc.), the HCM has often posed as the model used to compile other countries' own guidelines. In Europe in turn, the German (HBS) [5] can be considered the project guidelines model, therefore the authors took into account both manuals, as well as the guidelines used in Great Britain [6], Australia [7], Canada [8], and the Polish manual [9]. All of those dissertations contain calculation methods of determining traffic conditions at signalized intersections including methods for estimation of average vehicle queue lengths (average back-of-queue size).

A distinctive feature in the above mentioned guidelines [4–9] is the fact that for intersections with traffic lights they introduce methods for calculating the discharge time and the analyses of traffic conditions for substantial (hourly) values of traffic volumes (characteristic to the estimation of the functional efficiency of an intersection in the assumed analysis period). In planning and project applications, the typical, recommended analyses period is the period of 15 min - corresponding to the highest value of 15-min traffic volumes - due to significant changes in traffic fluctuations in traffic peak periods. It is advised to use the period of 1 h exclusively in cases when the changes in traffic volumes of vehicles arriving at an intersection inlet in 15-min sub periods are insignificant. Regardless of the choice of the analysis period, substantial measures of traffic conditions are first and foremost the average time losses, conditioning the later determination of the so called "Level of Service". This measure is parameterized according to the interval scale of the average time losses while driving through an intersection, connected with both the stopping of a vehicle during the red light display and the possibility of intersection inlet overcharging (when the duration of the green light display is not enough to serve the vehicles at the inlet). Average vehicle queue lengths, corresponding with such states (maximum and remaining queue respectively) [5–9], are determined similarly. With respect to vehicle queue lengths only the American guidelines [4], updated in 2010 and 2016, contain a calculation model which includes the so called initial queues (occurring at the intersection inlet at the beginning of the analysis period) in the course of calculation.

With regards to the course of calculation, a procedure including the period of 1 h, taking into account four 15-min sub periods (multiple interval analysis), is a more time consuming one. This procedure includes "overcharge shifting" marked in the given sampling interval between the individual analyses sub periods. This overcharge, identified with the "initial queue" is a simple balance of the demand surplus (traffic volume for vehicles arriving at the intersection inlet) over the supply (service ability - capacity of the intersection). This simple procedure is found in both the American guidelines [4], as well as for example in the Polish guidelines [9] - and it is recommended during a detailed diagnosis of the existing state of an intersection. It needs to be stressed here that in multiple interval analyses, the estimation of initial queue lengths is only used when determining average time losses and not in the method for determining the average length of remaining and maximum queues.

3 The Analysis of Vehicle Queue Lengths Research Results

3.1 The Analysis of Vehicle Queue Lengths Time Variability in Inlet Oversaturation

The consequence of a generalization of reliable traffic volume results for 1-h or 15-min periods is obtaining "very average" traffic conditions that existed or can exist in a particular element of a road infrastructure. An average value omits extreme states occurring in congestion periods (but not resulting from random incidents).

In order to present this better, an example will be described. It can be assumed that in a 1-h analysis period the density value does not exceed the capacity of the intersection in a 1-h period (it is about 10% lower). It will also be assumed that the analysis concerns a peak period (commuting from work or school, travelling to shopping centres, etc.). Simple calculations therefore show that there were no traffic conditions that would significantly deteriorate the functioning efficiency of the analyzed intersection inlet and that within that period all the vehicles arriving at the inlet should be served. It can also be assumed that hypothetically shortening the analysis period to 15 min will not cause significant fluctuations of the traffic volumes - it will be assumed that they are close to each other, and the fluctuations of the traffic density values do not exceed 5%. Intuitively it can be therefore assumed that there were no overcharges at that element of the road infrastructure.

The situation becomes highly complicated when analyzing fluctuations of the traffic volume in even shorter sampling intervals. Therefore, let one take into consideration the current conditioning of transport systems in large and medium-sized cities, when one is concerned with an increase in obligatory travel in the entire city area, practically in the very same moment. A situation of a very high transport demand is caused here, therefore a very large impulse of traffic volumes occurs (for example in a $\sim 5 \div 10$-min period), which causes significant disruptions in the traffic and leading to traffic overcharges at the inlets of many intersections. It is to be expected that the traffic density values for vehicles arriving at the inlet of a given intersection may exceed the ability to serve the vehicles during the green light display (the capacity) even several times in those several minutes. This means that the length of vehicle queues may reach beyond the designated accumulation areas, stretching even to the neighboring intersections (and such situations are encountered in reality after all). Should the traffic volume be kept on a similar level in the remaining sampling intervals (no significant fluctuations), permanent oversaturation of an inlet will occur. Only when the traffic volume will decrease in its value significantly (within several consecutive traffic signal cycles), will the vehicle queue be unloaded.

In order to better present the described example, the results of a research on the maximum queue lengths variability can be used, pictured in Fig. 2.

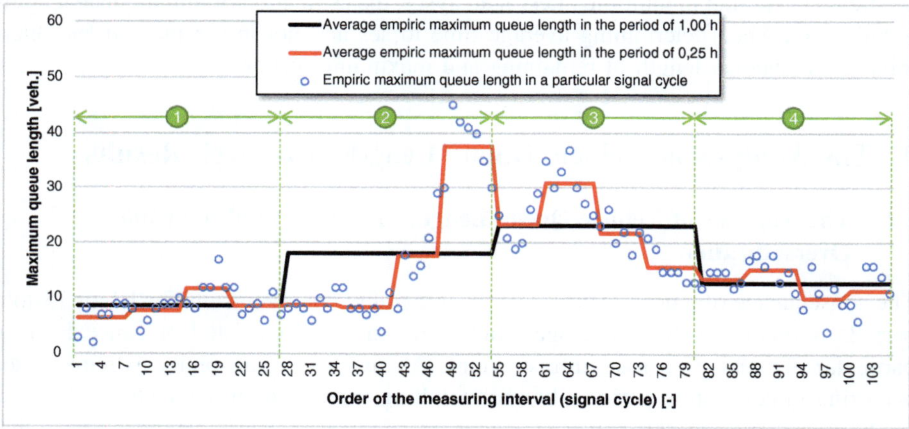

Fig. 2. Example results of a research on the maximum queue lengths variability from a 4-hour period (where: 1, 2, 3, 4 - measuring hour; measuring period 6:00÷10:00)

Using synthetic analysis on the research results presented in Fig. 2, it is clearly observed that in the period of the first and the fourth measuring hour, the average maximum queue length, in the analysis period of both 15 and 60 min, provides small estimation errors. In the case of the two remaining measuring hours (time period 7:00÷9:00), during which there was a communication peak, the estimated values for both sampling periods diverged from the empiric maximum lengths of vehicle queues, including of course large discrepancies in the case of a 1-h interval. Running the analysis in a shorter sampling interval, for example 5 min, adequately estimating the average values may capture the course of change in the process for both the increase and the unloading of the maximum queue, at the same time providing acceptable errors of the empiric average empiric vehicle queue lengths.

3.2 The Analysis of Errors in the Estimation of Vehicle Queue Lengths for Various Analysis Periods

The following time t_a frames were assumed to analyse the errors in the estimation of vehicle queue lengths: 1.00 h, 0.50 h, 0.25 h and 0.08 h. The difference between the value measured in a particular signal cycle and the average value defined from the empiric values in particular signal cycles in a given analysis period t_a was accepted as the estimation error.

The results of the analysis from all the research intervals were compiled in Table 1. Figure 3 illustrates the dispersion in the values of average maximum queues lengths as compared to the lengths of those queues occurring in the individual cycles for the above mentioned analysis periods, obtained from the research in one of the research intervals. And Fig. 4 shows the distributions of the values of errors in the estimation of average maximum queue lengths for those analysis periods.

Table 1. The results of statistical analysis of errors in the estimation of average queue lengths.

Queue type	t_a	Δk	θ	R
K_C	1.00	1.98	2.54	17.14
	0.50	1.83	2.38	16.23
	0.25	1.67	2.15	15.95
	0.08	1.34	1.75	13.33
K_P	1.00	3.63	5.55	54.66
	0.50	4.20	6.00	53.60
	0.25	3.68	5.37	49.00
	0.08	3.12	4.83	28.50
K_M	1.00	4.77	6.63	58.48
	0.50	3.71	5.03	50.60
	0.25	2.95	3.86	43.00
	0.08	1.89	2.45	19.67

where:
ta - duration of the analysis period [h];
Δx - average estimation error [veh.];
θ - standard deviation of error distribution [veh.];
R - error distribution gap [veh.].

The analyses results presented in Table 1 and in Figs. 1 and 2. Clearly show an increasing dispersion of the errors made while estimating the average values of vehicle queue lengths together with the increase of the analyses period. Despite the fact that the obtained average values of those errors are not high ($\sim 2 \div \sim 5$ veh.), special attention needs to be paid to the obtained dispersions of the estimated values, which are even up to $\pm$ 30 veh. When it comes to the maximum queue, the frequency of errors occurrence in the estimation of the relative value of vehicle queue lengths with the value higher than 5 veh. equals: $\sim$34% for a period of 1 h, $\sim$13% for a period of half an hour,

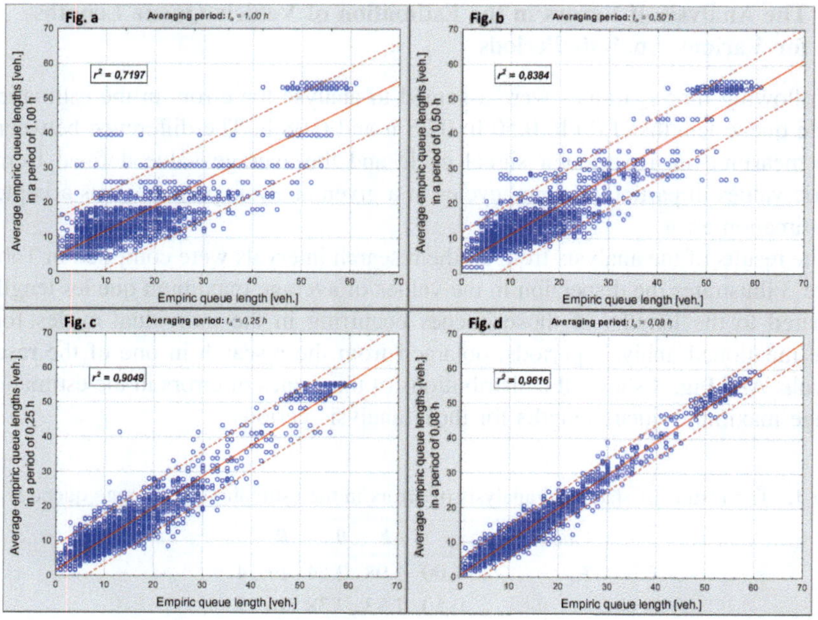

Fig. 3. The dispersion of the values of average empiric maximum queues as compared to the maximal queues occurring in the individual cycles for four analysis periods (Fig. a - 1.00 h, Fig. b - 0.50 h, Fig. c - 0.25 h, Fig. d - 0.08 h)

~11% for a period of 15 min, and ~3% for a period of 5 min. The quoted data indicate how large the scale of errors is when analyzing the traffic conditions for analyses periods exceeding the 5-min measuring interval.

It also needs to be noted that the average remaining queue lengths estimation errors have the main influence on the dispersion of the average empiric maximum queue lengths results, for the results of the analysis show that the smallest mean square error of average empiric values estimation concerns lengths of queues forming at red light display.

3.3 The Comparison of Average Empiric Values with Theoretical Values

In order to better show the currently occurring problem of imprecise vehicle queue lengths estimation at intersection inlets with traffic lights in the world, the analyses results illustrating the differences between the average values of maximum queues obtained from both research and estimation based on selected calculation methods will also be presented [4–9].

A detailed comparison of some of the average vehicle queue lengths estimation methods was described by the authors in their dissertation [10]. All the analyzed calculation methods provide very similar estimation results. General differences between those methods in the obtained results did not exceed ~1 veh. It only needs to be noted that no significant estimation correction results have been acquired within the updates of the Australian method (2013) and the American method (2010 and 2016). Because of

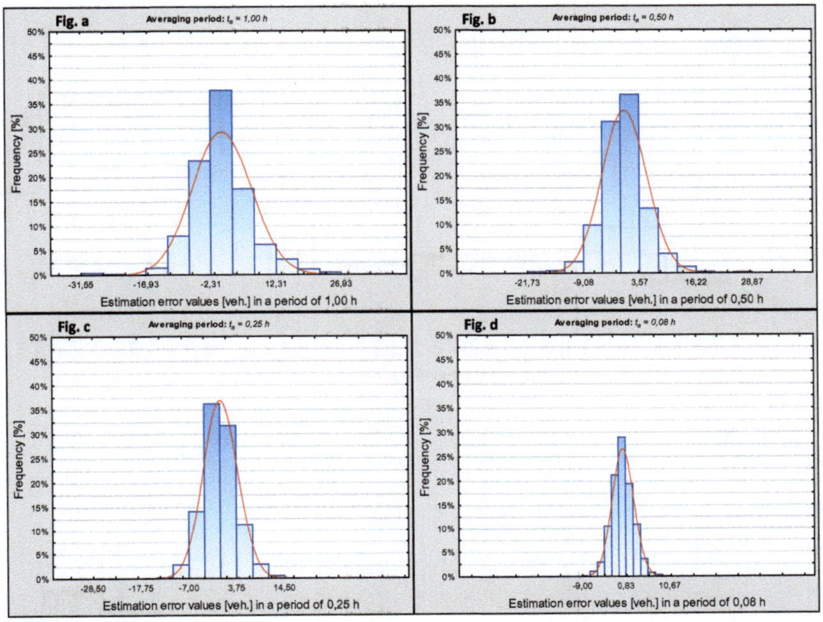

Fig. 4. The distribution of the error values in the estimation of the average empiric maximum queue lengths for four analysis periods cases (Fig. a - 1.00 h, Fig. b - 0.50 h, Fig. c - 0.25 h, Fig. d - 0.08 h)

that it was decided to perform simulation calculations for mathematical models included in the dissertations [4–9] and to establish an average value from the obtained results, which was then compared to with empiric values. The analyses results were compiled for sampling periods recommended in the guidelines, i.e. 1 h and 15 min.

This analysis was conducted for three inlets at three intersections with traffic lights. The results of the analysis are presented in Fig. 5. They clearly indicate an increase in average vehicle queue lengths estimation errors using theoretical methods in relation to average empiric values when assuming the sampling interval of 1 h in relation to a 15-min interval.

The analysis also shows a vast difference between empiric values and theoretical lengths of maximum queues for both analysis periods, for example around 245–265 on the first research section and 355–375 of a signal cycle (the periods of a very dynamic increase and decrease of the queue length).

In the case of the analyses of the average vehicle queue length theoretical values calculated using the quoted mathematical models [4–9], three conclusions need to be noted:

- estimating with their help of average vehicle queue lengths is rather parallel with the empiric values when no oversaturation of the intersection inlet occurs; when oversaturation does occur and it persists through a longer period (of several consecutive signal cycles), the estimated values do not reflect the traffic conditions prevailing at the intersection inlet whatsoever,

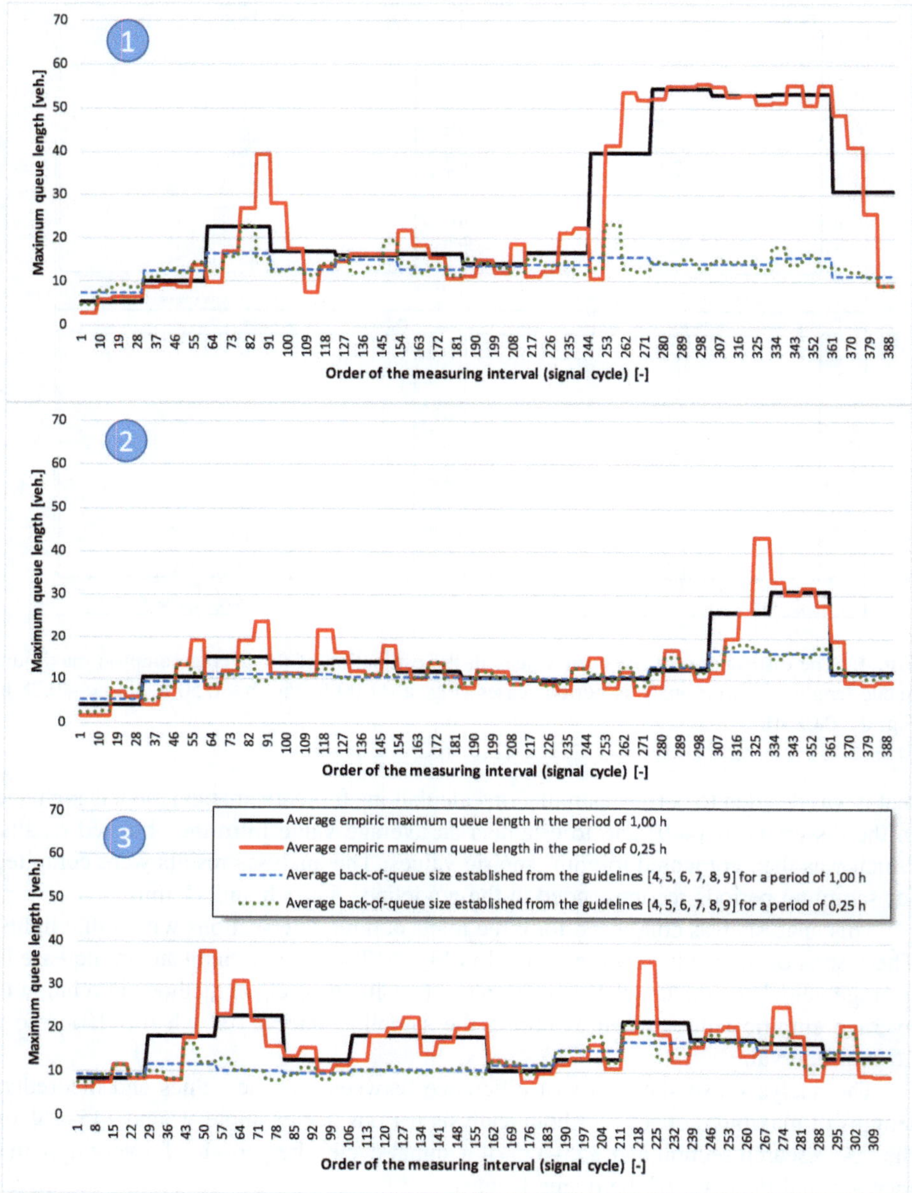

Fig. 5. The comparison of average empiric maximum queue length values with theoretical values for various durations of analyses periods (where: 1, 2, 3 - the number of a research section)

- as it was to be expected, the analyses results from the 0.25 h period are more accurate than the results of analyses conducted for the period of 1.00 h; however, it needs to be stressed that in the states of high traffic density at the inlets, the results

of queue lengths estimation are very inexact - regardless of the length of the chosen analysis period,
- the authors' earlier conclusions presented in the dissertation [10], concerning incorrect estimation of average vehicle queue lengths using the calculation methods employed worldwide in the states of increased crowding, are confirmed.

The end period from the upper chart's analysis (the first research section) is also worth mentioning. For almost 5 h of measuring, the queue length reached the neighboring intersection. As a result, the value of demand flow rate for vehicles arriving at the inlet of the analyzed intersection was approximately equal to the capacity of that inlet. Thus, technically, excluding the vehicles in the queue, the traffic conditions that prevailed corresponded with the inlet volume-to-capacity ratio at a 1:1 level. Owing to such status quo, the mathematical models estimated the average maximum queue length in a majorly understated way, while the empiric average queue length for both analysis periods was similar in that period and to a vast extent parallel with the results noted in the individual signal cycles.

4 Summary

The problem discussed in the article turns out to be a substantial one from the point of view of basic measurements of traffic conditions at intersections with traffic light, including vehicle queue lengths. It was clearly shown that with the increase of the analyses periods, the average estimation errors become larger, and so do the discrepancies in the estimation results.

The article also proves that averaging of the empiric vehicle queue lengths in periods longer than 15 min significantly "blurs" the actual traffic conditions at the inlets of signalized intersections (practically omitting extreme conditions whatsoever). In the authors' opinion, the problem of overcrowding of road networks poses a serious challenge in the nearest future for engineers and scientists, especially in developing countries, where the motorization indicator is constantly increasing (for example in Poland). Hence the authors' suggestion not to use a period of 1 h in the traffic conditions analyses whatsoever. The maximum period of traffic conditions analyses at intersections with traffic lights can be a period of 15 min, and a recommended one should be 5-min period.

Special attention should be paid to significant errors in average vehicle lengths estimation, based on calculation methods currently used in the world [4–9] in the states of overcrowding. Thus, establishing a correct calculation procedure becomes an urgent issue. Thorough establishing of remaining and maximum vehicle queue lengths at intersection inlets with traffic lights should contribute to more precise determination of traffic freedom degree, and in turn lead to optimal solutions from the scope of planning and designing of road infrastructure, constructing traffic signal control programs, or dynamic traffic management in urban areas.

References

1. European Commission: EU Transport in Figures. Statistical Pocketbook 2016. European Commission, Luxemburg (2016)
2. U.S. Department of Transportation: Passenger Travel. Fact and Figures 2016. Bureau of Transportation Statistics, Washington (2016)
3. Central Statistical Office: Transport - Activity Results in 2015. Central Statistical Office, Warsaw (2016)
4. Transportation Research Board: High Capacity Manual, Sixth Edition: A Guide for Multimodal Mobility Analysis. Transportation Research Board, Washington (2016)
5. Forschungsgesellschaft für Straßen und Verkehrswesen e. V.: Handbuch für die Bemessung von Straßenverkehrsanlagen 2015. Forschungsgesellschaft für Straßen und Verkehrswesen, Köln (2015)
6. The Highway Agency, Scottish Executive Development Department, Welsh Assembly Government, The Department for Regional Development Northern Ireland: Design Manual for Roads and Bridges. Volume 6 - Road Geometry. Section 2 - Junctions. Wiliamsleatag, England (2017)
7. Austroads: Guide to Traffic Management. Part 3: Traffic Studies and Analysis. Austroads, Sydney (2013)
8. The Institute of Transportation Engineers: Canadian Capacity Guide for Signalized Intersections, 3rd edn. The Institute of Transportation Engineers, Ottawa (2008)
9. General Directorate for National Roads and Motorways: A Method of Calculating the Capacity of Intersections with Traffic Lights. General Directorate for National Roads and Motorways, Warsaw (2004)
10. Szczuraszek, T., Iwanowicz, D.: The analysis of calculation methods of vehicle queue lengths in front of intersection inlets with traffic lights. Logistyka **4**, 1523–1534 (2015)

Comparative Analysis of Chosen Adaptive Traffic Control Algorithms

Krzysztof Małecki(✉) and Piotr Pietruszka

Faculty of Computer Science,
West Pomeranian University of Technology, Szczecin, Poland
{kmalecki,ppietruszka}@wi.zut.edu.pl

Abstract. Setting optimal scenario for traffic lights on crossroads is very important task related to modeling of modern, ordered traffic in smart cities. In this article modifications of traffic lights phases control algorithms on crossroads with different densities have been presented. Comparative analysis of chosen algorithms effectiveness for defined area has been also made. Particularly strategies based on traffic detectors placed in front or behind a crossroad and algorithm "injecting" cars have been compared. The second solution is dedicated to situation with autonomous (driverless) vehicles working with high time accuracy. Solutions developed using traffic simulation allowed to prove that proposed modifications of traffic lights control algorithms can improve effectiveness in specified cases.

Keywords: Intelligent · Transportation systems · Traffic efficiency · Traffic simulators · Adaptive traffic signal control

1 Introduction

The essential problem in cities is growing traffic density. The system effectiveness improvement is possible to achieve by road network development or by improving traffic patency. This subject, known as effective traffic management, including for example traffic lights control, is a part of Intelligent Transportation Systems (ITS) [1]. Nowadays there are three main strategies of traffic lights control: constant in time, actuated and adaptive [2].

When establishing the schedule for constant time based traffic lights, historical data on traffic in different time of the day is being used. However the settings are not always suitable for current needs. To deal with real time traffic lights control, adaptive strategies were introduced. They usually base on signals from parts of road infrastructure called detectors – video cameras, induction loops, etc. The well known adaptive strategies are: ACS-Lite [3], Scats [4], Scoot [5], OPAC [6], UTOPIA [7].

The goal of this article is to analyze and compare efficiency of selected algorithms and proposal of change in existing traffic light control algorithms. Literature analysis was made; the area of simulated environment was defined and evaluated chosen strategies and algorithms performance.

E. Macioszek and G. Sierpiński (eds.), *Recent Advances in Traffic Engineering for Transport Networks and Systems*, Lecture Notes in Networks and Systems 21,
DOI 10.1007/978-3-319-64084-6_18

2 Related Work

The literature presents different methods and technologies, which allow managing traffic lights more efficiently. Generally it can be said that the essence of controlling traffic lights are algorithms switch traffic lights depending on traffic density [8]. However it should be noted that it is very common to use sensors such as induction loops and video cameras to collect data and pass it to algorithms dedicated to traffic light control. In [9] authors suggested using Simultaneous Perturbation Stochastic Algorithm (SPSA) to compute in real time the best setting of signals in order to minimize the time of passing the crossroad. Some research concerns about signal control using video cameras. In [10] authors uses video stream from cameras on the crossroad to compute current traffic density in real time. Obtained results feed the algorithm switching traffic lights according to cars density, which is supposed to decrease traffic jams, thereby decreasing number of accidents. In [11] authors developed new controller (EOM-ANN Controller) of traffic lights based on neural networks. In this area of research, also the idea of self-organizing traffic lights has been described. In [12] authors presented distributed algorithm for traffic lights control. Hybrid algorithms [13] that combines Fuzzy Logic Controller (FLC) and Genetic Algorithms (GA) have been developed and tested for effectiveness. Distributed scheduling strategy based on Lagrangian relaxation and subgradient method have been presented in [14]. The next proposal is applying warning lights supporting traffic lights to avoid traffic jams [15]. As in constant time based traffic lights, as well in adaptive genetic algorithms are used. In this case they are not used to set duration of each phase, but for example for finding weights to apply in artificial neural network [16]. The issues of traffic control algorithms and modelling were also discussed in [17–26].

3 Theoretical Background

This section describes three approaches for optimization of adaptive traffic lights. All the solutions are similar, in the idea to activities, to the policeman controlling traffic at the intersection and drawing attention to the length of the vehicles queue waiting to enter the intersection or lack of place after the crossroad.

3.1 Evaluating the Length of Queue Before the Crossroad

Authors of [27] focus on providing optimal local solutions, which means that it is going to work in the area of one crossroad. Traffic light phase duration adjustment algorithm was shown on example of road network, which consist only of one-way roads from south to north and from west to east and distances between all adjacent crossroads are equal. Algorithm works for traffic lights cycle containing two phases and whole cycle duration, which is the sum of both phases' durations, is constant. First step of the algorithm is the initialization of phase's duration, which is set to half of cycle duration. Afterwards next steps of simulation are being run in a loop and in each iteration queue length in each direction is being kept in memory. Alter some time (called decision time), in case of queue lengths in intersecting directions differing by more than 10%, phases duration are being changed as follows: it is being increased for direction with longer queue and decreased for perpendicular direction.

3.2 Using Induction Loops Behind Crossroads

Research published in [28] concerns about placing induction loops behind the crossroad to measure cars density there and make green light phase shorter if there is no possibility to drive after passing the crossroad. Different options of placing induction loops behind the crossroad have been tested using cellular automata based simulator. Presented solutions have been compared to the situation which could occur without switching traffic lights state basing on data collected from sensors in both cases: when drivers obey the law in term of not entering the crossroad when there is no place behind to leave it and when they break this rule. When the traffic was intense, higher throughput was achieved on crossroads with induction loops.

3.3 CarInjection

The main assumption of the CarInjection algorithm [29] was to allow passing the crossroad by the cars, when there are no cars approaching on perpendicular direction. When traffic density on one of intersecting roads, for which traffic light is green, is so low that there is a gap between cars (which means that there is a free space of given length between traffic light and next approaching car), the algorithm forces earlier lights change. This behavior leads to allowing the cars waiting on red light on perpendicular road to receive green light and pass the crossroad, and the signal for road with free space becomes red. In case of high traffic density, when it is not possible to find free space on any road, lights are controlled by previously set constant time program.

4 Research Finding

In this section the research environment and simulation scenario are presented. Then effectiveness measures for comparing algorithms are defined.

4.1 Research Environment

To do the research SUMO (Simulation of Urban MObility) [30] traffic simulator was used. The main reason was the popularity of SUMO in similar research, for example in [31–34]. To do the research described in this article the software written in Python language was developed. It was integrated with SUMO using interface named TraCI (Traffic Control Interface) [35], which allows getting information about current state of objects and changing their state. It's capabilities include for example getting data from detectors, getting vehicles data, getting and setting state of traffic lights.

SUMO allows placing on roads three different traffic detectors: induction loops, lane area detectors and multi-entry multi-exit detectors.

4.2 Simulation Scenario and Effectiveness Measures

To test the effectiveness of traffic lights control algorithms, the simulation scenario containing the map of roads network (Fig. 1), placement of detectors and vehicles generation setup has been developed, according to [36]. Traffic lights on each intersection is working using two-phase program: the first phase allows movement of vehicles on roads from north to south and from south to north, and the second phase – from west to ease and from east to west.

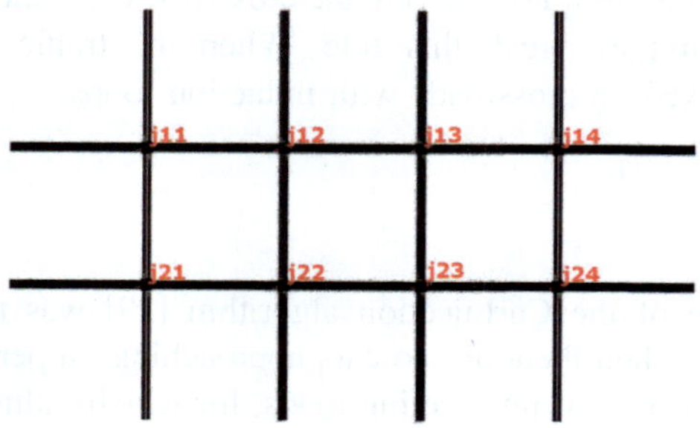

Fig. 1. Screenshot of map of the tested area opened in SUMO with crossroads identifiers added

Vehicles movement has been set up to ensure appropriate traffic density, depending on configuration file that has been used. Depending on needs, we have used following types of files with density configuration: cars can drive straight or turn right once; cars can only drive straight; cars can drive straight and density is constant and has a high value for west-east direction, but density for north-south direction is being set for each configuration file. The effectiveness of given algorithm can be measured using following criteria: throughput, probability of congestion relieve, etc.

In this work two criteria have been used to compare tested algorithms: number of vehicles which drove through the whole network of roads and accumulated time when cars are stopped in front of traffic light or another car.

5 Test Methods of Algorithms for Adaptive Traffic Lights

To test the effectiveness of presented solutions for adaptive traffic lights, traffic simulations using each of presented methods have been run with different traffic densities. Data on the total time to stop vehicles are the average of a number of simulations for each case. The final simulation results are the average of the results for the total time to stop vehicles compared with the corresponding average of the results obtained using fixed traffic lights. Duration of each simulation has been set up to 1500 steps.

Algorithm of phase duration modification has been adjusted to simulation scenario in the following way: queue length on one-way road has been replaced with sum of queues lengths on two opposite lanes of two-way road. Initial phases durations for each traffic light have been set to values obtained as a result of constant time based traffic lights optimization. The threshold value of difference between queues lengths have been set to 10%, exactly as in original solution and time between making decisions have been set to 100 steps instead of 300 to make it possible for control algorithm to adjust traffic lights phases duration to traffic density before the end of whole simulation (1500 steps).

To determine length of queue, lane area detectors available in SUMO have been used. They have been placed on every entry of every crossroad. Length of each detector has been set to 50 meters. Placement of detectors for example of one crossroad is shown on Fig. 2a.

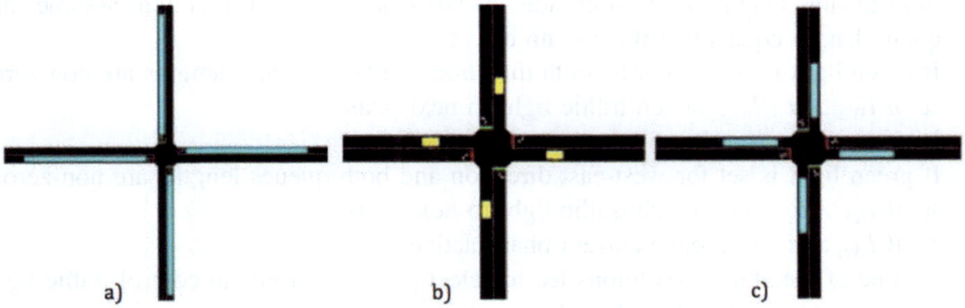

Fig. 2. Arrangement of the road detectors (a) lane area at the crossroad - in the solution with the modification of the length of the phases based on the queue length (b) induction loops; (c) area detectors

From presented cases of placing detectors behind the crossroad, the following two strategies have been chosen:

1. setting red light for direction, for which stopped cars were detected on lanes in both sides – to allow this solution to work properly, inductions loops have been moved further from crossroad comparing to original solution (they are placed not directly behind crossroad, but about 10 meters from it) – Fig. 2b,
2. setting red light for direction, for which queue behind crossroad is longer (comparison is made between sum of queues lengths on both opposite sides of crossroad) – Fig. 2c.

The CarInjection algorithm has been implemented according to description in Theoretical background section. The condition of existence of free space in front of crossroad is fulfilled, if on values gathered from detectors on both sides of crossroad allows finding out that there is no car in this place. To detect free space, lane area detectors have been used. They have been places as shown on Fig. 2a, but the length has been limited to 20 m. Detectors length have been chosen basing on several experiments, which showed that if the car approaching the crossroad is more than 20 m away, then it can receive red light to give way to car from perpendicular direction and traffic light will switch back to green until that car stops. We assume that car is autonomous and there is not human factor (the driver), whose reaction can include changing of vehicle speed.

5.1 New Solutions Proposals

Two new solutions for choosing current traffic light phase, based on solutions described above, are developed. They use values gathered from lane area detectors.

Determining length of queue before crossroad - method based on [17], but in this case, algorithm is controlling traffic light "live", not using data gathered in several hundred of simulation steps. This can cause faster reaction of controller to changing traffic density. Algorithm presented as list of steps:

1. Sum queues lengths from both sides of crossroad for each direction, assume that queue length equals 0 if there is no car.
2. If green light is set for north-south direction and both queues lengths are non-zero:
   a. if $l_{WE} > \alpha \cdot l_{NS}$, switch traffic light to next phase,
   b. if $l_{NS} > \alpha \cdot l_{WE}$, leave current phase active.
3. If green light is set for west-east direction and both queues lengths are non-zero:
   a. if $l_{NS} > \alpha \cdot l_{WE}$, switch traffic light to next phase,
   b. if $l_{WE} > \alpha \cdot l_{NS}$, leave current phase active.
4. If none of the above conditions led to select phase, continue to control traffic light according to constant time based program.

In formulas in algorithm description the following symbols have been used: l_{NS} - sum of queue lengths for north-south direction, l_{WE}- sum of queue lenghts for west-east direction, α - constant factor.

This method has been tested for different values of factor α, which goal is to decrease sensitivity of controller, to disallow too frequent phase changes.

Occupied road ratio indicated by lane area detectors - the second strategy is based on ration of occupied space gathered from lane area detectors (what part of detector area is occupied by vehicles). It consist of following steps:

1. Compute sum of occupation ratio on both crossroad sides for each direction.
2. If green light is set for north-south direction:
   a. if $o_{WE} > \beta \cdot o_{NS}$, switch traffic light to next phase,
   b. if $o_{NS} > \beta \cdot o_{WE}$, leave current phase active.
3. If green light is set for west-east direction:
   a. if $o_{NS} > \beta \cdot o_{WE}$, switch traffic light to next phase,
   b. if $o_{WE} > \beta \cdot o_{NS}$, leave current phase active.
4. If none of the above conditions led to select phase, continue to control traffic light according to constant time based program.

The following symbols have been used: o_{NS} - sum of occupation ratio in front of crossroad on north-south direction; o_{WE} - sum of occupation ratio in front of crossroad on west-east direction; β - constant factor. Using these factors disallows changing lights too often, what has been identified as a problem on the beginning of this algorithms development.

6 Research Results

Values presented in tables are results obtained when testing given algorithm as a percent of result of constant time based traffic lights control. The 100 value means there is no change, higher value means worse result and lower value means better result. Algorithms in tables are marked as follows: q_l – phase duration change based on queue length in defined time; b_1, b_2- application of induction loops behind crossroads; c_i – CarInjection algorithm; j_l – method based on queue length in front of crossroad, with α factor presented in table; o_{cc} – method based on space occupancy in front of crossroad, β factor values are presented in Table 1.

Table 1. Changing the sum of times to stop vehicles using different algorithms for the configuration of the defined values of the traffic (identical for all roads) that have the option turn right.

Algorithm	Intensity									
	0.010	0.064	0.119	0.173	0.228	0.282	0.337	0.391	0.446	0.500
q_l	98.15	475.2	158.33	298.11	348.36	540.84	239.35	343.85	194.33	134.44
b_1	100	100.27	216.56	69.85	99.63	97.88	100.05	99.52	100.51	99.96
b_2	100	1597	1556	784.58	1493	1672	1673	1426	1476	1566
c_i	0.05	101.81	92.14	127.73	314.91	83.59	95.71	88.43	227.1	92.96
$j_l(\alpha = 8)$	100	91.91	115.36	50.37	90.17	86.31	91.58	94.29	94.4	96.59
$j_l(\alpha = 11)$	100	96.1	91.94	80.98	87.36	184.07	96.29	99.99	99.38	97.47
$j_l(\alpha = 14)$	100	98.54	93.27	52.43	207.59	91.68	99.97	100.93	100.6	101.12
$o_{cc}(\beta = 3)$	1.83	92.43	574.72	46.86	116.09	132.71	90.05	204.44	85.2	250.13
$o_{cc}(\beta = 4.3)$	1.62	120.01	239.98	150.06	85.77	81.26	89.6	100.08	176.77	302.83
$o_{cc}(\beta = 5.6)$	1.55	87.28	222.47	175.29	88.13	99.68	89.51	91.78	210.68	109.78

The best solution when traffic density is very low is CarInjection algorithm, It is caused by the case when density is low, car approaching the crossroad on one road can cause immediate switching of its direction light to green, because it is very unlikely that there is a car approaching this intersection on the perpendicular direction at the same time.

The worst result has been achieved for one of methods using detectors placed behind crossroad. The cause of it is that algorithm is checking what is happening behind the crossroad and if there is a jam for cars driving strait through the crossroad, it switches traffic light phase. If the car waiting in front of traffic light has set route to turn right on this crossroad and it receives green light, the traffic jam on perpendicular causes blocking cars on this intersection (Table 2).

For each of tested configurations, it turns out that the best methods are CarInjection and algorithm using occupied space evaluation. First of them is better for lower traffic densities, but the second is proved to be more effective fro higher densities. The rest of algorithms gave us better results in some cases, but it was not meaningful and

Table 2. Changing the sum of times to stop vehicles using different algorithms for the configuration of the defined traffic values (identical for all roads), passing straight through crossroad.

Algorithm	Intensity									
	0.010	0.064	0.119	0.173	0.228	0.282	0.337	0.391	0.446	0.500
q_l	94.26	84.89	87.55	88.17	86.96	88.55	88.32	88.58	88.49	88.21
b_1	100	100	100.18	99.65	100.08	100.12	99.77	100.01	99.93	100.11
b_2	100	100.97	91.65	89.49	89.84	90.98	90.51	91.2	90.43	91.22
c_i	0.38	3.63	74.55	82.75	82.01	85.48	83.88	82.37	84.87	84.64
$j_l(\alpha = 8)$	100	100.16	90.34	88.61	87.83	88.37	88.9	88.4	89	88.68
$j_l(\alpha = 11)$	100	100	96.54	94.6	94.39	96.88	96	95.41	95.05	94.66
$j_l(\alpha = 14)$	100	100	98.36	98.1	98.38	97.93	98.8	97.91	98.05	97.36
$o_{cc}(\beta = 3)$	0.35	11.4	54.65	60.73	60.65	61.03	60.44	62.97	61.81	61.96
$o_{cc}(\beta = 4.3)$	0.32	14.04	64.52	73.52	71.96	71.84	73.72	73.06	72.77	73.88
$o_{cc}(\beta = 5.6)$	0.32	16.74	69.63	77.48	77.57	77.85	79.81	77.97	79.79	79.27

sometimes results were significantly worse. Worst results have been observed for application of detectors behind crossroads, but in case of high density of cars moving straight through crossroad it showed some effectiveness improvement, which lets us to say that this solution could be used not instead other algorithms, but as a method which supports them.

7 Summary

This article focuses on current problem of modelling effective traffic lights control. The goal of the research was to propose changes in adaptive traffic light control algorithms and tries to achieve chosen algorithms effectiveness. Two algorithms changes have been proposed and tested using simulator named SUMO extended with software developed during this research. Methods based on comparing selected data gathered from detectors placed on every crossroad on entries and exits have been used. Decisions have been made based on cars existence and speed data collected from induction loops, as well as queues lengths and space occupancy collected from lane area detectors available in simulator. Conducted Experiments proved that developing algorithm, which would be common for different traffic types is a very hard task to solve. However during the research, best results have been achieved using CarInjection algorithm and one of proposed strategies, which based on space in front of crossroad occupancy.

References

1. Taniguchi, E., Shimamoto, H.: Intelligent transportation system based dynamic vehicle routing and scheduling with variable travel times. Transp. Res. Part C: Emerg. Technol. **12** (3), 235–250 (2004)

2. Feng, Y.H., Head, K.L., Khoshmagham, S., Zamanipour, M.: A real-time adaptive signal control in a connected vehicle environment. Transp. Res. Part C **55**, 460–473 (2015)
3. Luyanda, F., Gettman, D., Head, L., Shelby, S., Bullock, D., Mirchandani, P.: ACS-lite algorithmic architecture: applying adaptive control system technology to closed-loop traffic signal control systems. Transp. Res. Rec. **1856**, 175–184 (2003)
4. Sims, A.G., Dobinson, K.W.: The Sydney coordinated adaptive traffic (SCAT) system philosophy and benefits. IEEE Trans. Veh. Technol. **29**(2), 130–137 (1980)
5. Hunt, P.B., Robertson, D.I., Bretherton, R.D., Royle, M.C.: The SCOOT on-line traffic signal optimisation technique. Traffic Eng. Control **23**(4), 190–199 (1982)
6. Gartner, N.: OPAC: a demand-responsive strategy for traffic signal control. Transp. Res. Rec. **906**, 75–81 (1983)
7. Mauro, V., Taranto, C.D.: UTOPIA. IFAC Control, France (1989)
8. Roozemond, D.A.: Using intelligent agents for pro-active, real-time urban intersection control. Eur. J. Oper. Res. **131**(2), 293–301 (2001)
9. Kwatirayo, S., Almhana, J., Liu, Z., Siblini, J.: Optimizing road intersection traffic flow using stochastic and heuristic algorithms. In: IEEE International Conference on Communications 2014, pp. 586–591. IEEE Press, Sydney (2014)
10. Kanungo, A., Sharma, A., Singla, C.: Smart traffic lights switching and traffic density calculation using video processing. In: Proceedings of 2014 Recent Advances in Engineering and Computational Sciences, pp. 1–7. IEEE Press, Chandigarh (2014)
11. De Oliveira, M.B.W., De Almeida, N.A.: Optimization of traffic lights timing based on artificial neural networks. In: 17th IEEE International Conference on Intelligent Transportation Systems, pp. 1921–1922. IEEE Press, Qingdao (2014)
12. Goel, S., Bush, S.F., Ravindranathan, K.: Self-organization of traffic lights for minimizing vehicle delay. In: International Conference on Connected Vehicles and Expo, pp. 931–936. IEEE Press, Vienna (2015)
13. Odeh, S.M.: Hybrid algorithm: fuzzy logic-genetic algorithm on traffic light intelligent system. In: IEEE International Conference on Fuzzy Systems, pp. 1–7. IEEE Press, Istambul (2015)
14. Zhang, Y., Su, R., Gao, K.: Urban road traffic light real-time scheduling. In: Proceedings of the IEEE Conference on Decision and Control, pp. 2810–2815. IEEE Press, Osaka (2016)
15. Qi, L., Zhou, M., Luan, W.: An Emergency Traffic Light Strategy to Prevent Traffic Congestion. 13th IEEE International Conference on Networking. Sensing and Control, pp. 1–6. IEEE Press, Mexico City (2016)
16. Ruchaj, M., Stanisławski, R. Approaches to control acyclic traffic lights in an exemplary urban road network. Methods and models in automation and robotics. In: Proceedings of the 16th International Conference on Methods and Models in Automation and Robotics, pp. 387–392. IEEE Press, Międzyzdroje (2011)
17. Oskarbski, J., Gumińska, L., Miszewski, M., et al.: Analysis of Signalized Intersections in the Context of Pedestrian Traffic. Transp. Res. Proc. **14**, 2138–2147 (2016)
18. Oskarbski, J., Miszewski, M., Żarski, K.: Traffic control within bus queue jumps on the example of Gdynia. Transp. Miejski i Regionalny **4**, 38–43 (2015)
19. Bazan, M., Ciskowski, P., Halawa, K., Janiczek, T., Kozaczewski, P., Madej, Ł., Rusiecki, A.: Intelligent transport system auditing using road traffic micro-simulation. Arch. Transp. Syst. Telemat. **8**(4), 3–8 (2015)
20. Bazan, M., Ciskowski, P., Janiczek, T., Madej, Ł., Rusiecki, A., Halawa, K.: Green wave optimization. Arch. Transp. Syst. Telemat. **9**(3), 3–8 (2015)
21. Karoń, G., Mikulski, J.: Transportation systems modelling as planning, organisation and management for solutions created with ITS. In: Mikulski, J. (ed.) Modern Transport Telematics. CCIS, vol. 239, pp. 277–290. Springer, Berlin, Heidelberg (2011)

22. Karoń, G., Mikulski, J.: Problems of ITS architecture development and ITS architecture implementation in Upper-Silesian conurbation in Poland. In: Mikulski, J. (ed.) Telematics in the Transport Environment. CCIS, vol. 329, pp. 183–198. Springer, Berlin, Heidelberg (2012)
23. Karoń, G., Żochowska, R.: Modelling of expected traffic smoothness in urban transportation systems for ITS solutions. Arch. Transp. **33**(1), 33–45 (2015)
24. Żochowska, R., Karoń, G.: ITS Services packages as a tool for managing traffic congestion in cities. In: Sładkowski, A., Pamuła, W. (eds.) Intelligent Transportation Systems-Problems and Perspectives. SSDC, vol. 32, pp. 81–103. Springer, Switzerland (2016)
25. Sobota, A., Kłos, M., Karoń, G.: The influence of countdown timers on the traffic safety of pedestrians and vehicles at the signalized intersection. In: Sierpiński, G. (ed.) Intelligent Transport Systems and Travel Behaviour. AISC, vol. 505, pp. 13–21. Springer, Switzerland (2017)
26. Wnuk, D., Karoń, G.: Variants to improve traffic conditions in the bottleneck in road network of the Upper-Silesian conurbation. In: VI International Scientific Conference Transport Problems Conference Proceedings on CD-ROM, pp. 970–981. Silesian University of Technology, Katowice (2014)
27. Mikat, J., Brockfeld, E., Wagner, P.: Agent based traffic signals on a basic grid. In: Proceedings of the 4th Workshop on Agent-Based Simulation, pp. 1–6. SCS European Publishing House, Erlangen (2003)
28. Małecki, K., Pietruszka, P., Iwan, S.: Selected aspects of traffic micro-simulation based on cellular automata and traffic detection system. Arch. Transp. Syst. Telemat. **6**(1), 41–44 (2013)
29. Małecki, K.: The importance of automatic traffic lights time algorithms to reduce the negative impact of transport on the urban environment. Transp. Res. Proc. **16**, 329–342 (2016)
30. Krajzewicz, D., Hertkorn, G., Rössel, C., Wagner, P.: SUMO (Simulation of Urban MObility) - an open-source traffic simulation. In: Proceedings of the 4th Middle East Symposium on Simulation and Modelling, pp. 183–187. SCS European Publishing House, Erlangen (2002)
31. García-Nieto, J., Alba, E., Olivera, A.C.: Swarm intelligence for traffic light scheduling: application to real urban areas. Eng. Appl. Artif. Intell. **25**(2), 274–283 (2012)
32. Krajzewicz, D., Erdmann, J., Behrisch, M., Bieker, L.: Recent development and applications of SUMO - simulation of urban mobility. Int. J. Adv. Syst. Meas. **5**(3&4), 128–138 (2012)
33. Gu, W., Ito, T.: Optimization of road distribution for traffic system based on vehicle's priority. In: Booth, R., Zhang, M. (eds.) PRICAI 2016: Trends in Artificial Intelligence. LNCS, vol. 9810, pp. 729–737. Springer, Switzerland (2016)
34. Cárdenas-Benítez, N., Aquino-Santos, R., Magaña-Espinoza, P., Edwards-Block, A., Cass, A.M.: Traffic congestion detection system through connected vehicles and big data. Sensors **16**(5), 1–26 (2016)
35. SUMO Simulator. http://sumo.dlr.de/wiki/traci
36. Małecki, K., Pietruszka, P., Iwan, S.: Comparative analysis of selected algorithms in the process of optimization of traffic lights. In: Nguyen, N.T., Tojo, S., Nguyen, L.M., Trawiński, B. (eds.) Intelligent Information and Database Systems. LNAI, vol. 10192, pp. 497–506. Springer, Berlin (2017)

Traffic Time Delay Modelling on the Intersection in the City of Martin, Using Software Aimsun

Alica Kalašová(✉), Jana Kupčuljaková, Simona Kubíková, and Jozef Paľo

The Faculty of Operation and Economics of Transport and Communications, University of Žilina, Žilina, Slovakia
{alica.kalasova, jana.kupculjakova, simona.kubikova, jozef.palo}@fpedas.uniza.sk

Abstract. In recent years, there has been a large increase in individual transportation due to rising living standard and systematic economic preference of individual car transportation and a road network in terms of capacity and permeability ceases to satisfy. Traffic collapses, congestions, accidents and a further negative impact on the environment are gradually becoming a common part of our lives. In spite of large investments into the infrastructure development, the traffic situation is constantly getting worse. In our contribution, we analyze the intersection of roads n. I/65 and III/2144 near the city of Martin, which is very busy and we propose possible solutions using microscopic simulation in software Aimsun.

Keywords: Traffic survey · Basic capacity · Safety · Microscopic simulation

1 Introduction

Transportation is a phenomenon consists of a large amount of particular features, which create a comprehensive unit. Human activity is always primary within transportation because manages and controls movement of transport modes in time and space (except planning and organizing).

In recent years, there has been a large increase in individual transportation due to rising living standard and systematic economic preference of individual car transportation. Road network in terms of capacity and permeability ceases to satisfy. Traffic collapses, congestions, accidents and further negative impact on the environment are gradually becoming a common part of our lives. In this case, new communications and interchanges are building.

In spite of large investments into the infrastructure development, the traffic situation is constantly getting worse. An intersection is a point where traffic flows are meet, connect, disconnect and intersect from all intersecting roads. It is a point where in ground plan road communications meet or intersect and they are connected to each other. [1] Because of these facts, an intersection has to have sufficient permeability. Otherwise, all traffic flows could stagnate on the intersection or in front of it. When we are making decision about type, shape and location of an intersection we have to take into account the requirement to ensure safety and smoothness of traffic flows.

E. Macioszek and G. Sierpiński (eds.), *Recent Advances in Traffic Engineering for Transport Networks and Systems*, Lecture Notes in Networks and Systems 21,
DOI 10.1007/978-3-319-64084-6_19

2 Characteristic of the Chosen Intersection

The chosen intersection of roads I/65 and III/2144 is situated near the city of Martin, about 4 km of Bystrička village. Most of active working population commute to work to Martin. The chosen intersection is very problematic and the main road is overloaded. Figure 1 shows the intersection highlighted in the circle. The intersection entrance from Bystrička- left turn is very hard to drive through and also there are high delays on the entrance to the village especially in the morning peak hour.

Fig. 1. Location of the chosen intersection (Source [5])

2.1 Realization and Evaluation of Junction's Transport Survey

Our task was to analyze and solve delays on the intersection entrances D and C. We have performed a transport survey on the chosen intersection (Fig. 2). The survey was focused on traffic intensity on each enter to intersection and also routing of traffic. After that, we evaluated this transport survey, analyzed transport situation and identified possible solutions.

Junction counting is a special type of direction transport survey, where traffic routing is determine on small area. It is used to detect the movement of vehicles in the intersection, and thus to determine the intensity of traffic in each direction in the observed time interval. Accuracy of traffic peak hour depends on the chosen time interval. The previous surveys have shown that survey of intensity of individual transportation should be made in an average day of week (Tuesday, Wednesday or

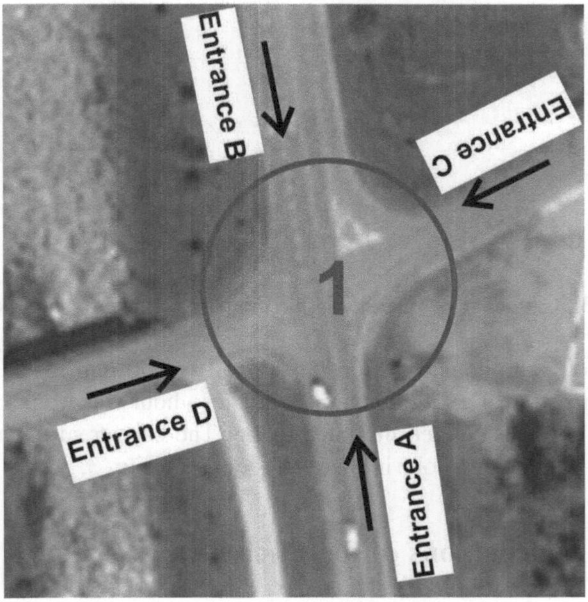

Fig. 2. Location of the chosen intersection and marking of standpoints (Source own study according [5])

Thursday) and in months from March to June, September and October. The minimum duration of survey should be 4 h.

We have performed this transport survey on Thursday 2nd of June 2016. The first part we were performing in time from 6:00 a.m. to 10:00 a.m. and the second one from 1:00 p.m. to 5:00 p.m.

Vehicles which entranced to junction we have recorded to census form in the chosen time intervals and divided according to type of vehicle and direction. Location of each standpoints are shown on Fig. 2.

2.2 Evaluation of Transport Survey

The total number of real vehicles, which entered the intersection, is:

- from 6:00 a.m. to 10:00 a.m. - 6 106 vehicles,
- from 1:00 p.m. to 5:00 p.m. - 6 701 vehicles,
- during 8 h of survey the sum of vehicles was 12 807.

The morning peak hour was from 7:15 a.m. to 8:15 a.m. when 1 724 vehicles entered intersection. The morning peak quarter-hour was from 7:45 a.m. to 8:00 a.m. when 493 vehicles entered the intersection. The cartogram of intensity on each entrance during traffic peak hour in the morning is on Fig. 3.

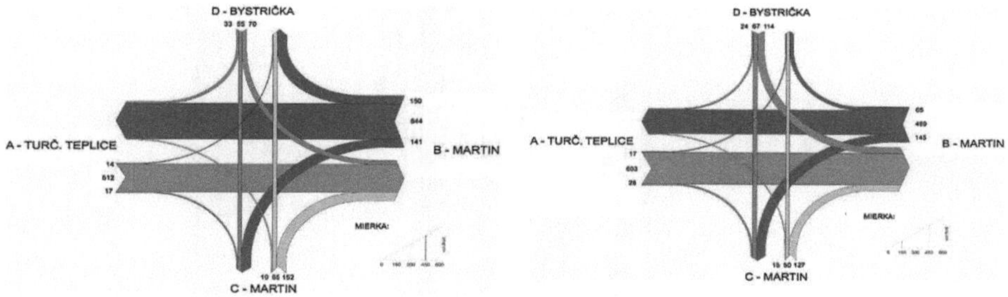

Fig. 3. Cartogram of intensity in the morning and afternoon peak hour

The afternoon peak hour was from 3:00 p.m. to 4:00 p.m. when 1 872 vehicles entered the intersection. The afternoon peak quarter - hour was from 3:45 p.m. to 4:00 a.m. when 520 vehicles entered the intersection. The cartogram of intensity on each entrance during traffic peak hour in the afternoon is on Fig. 3.

3 Capacity Calculations of the Chosen Intersection

We have used obtained information from transport survey to capacity calculations according to technical conditions TC 16/2015. Individual calculating procedures serve as a base for a documentation, which considers a fact that the chosen intersection permeability is sufficient for traffic volume with the movement in required quality. The intersection is uncontrolled. Calculations are based on data which were obtained from transport survey. The permeability of the intersection is approximately 600–800 vehicles per hour. Pedestrians and cyclists are not taking into account in calculations [2]. The actual traffic volume of all allowed traffic movements at the intersection is a prerequisite for using the following methods of calculation. It is necessary to define levels of traffic flows (Fig. 4).

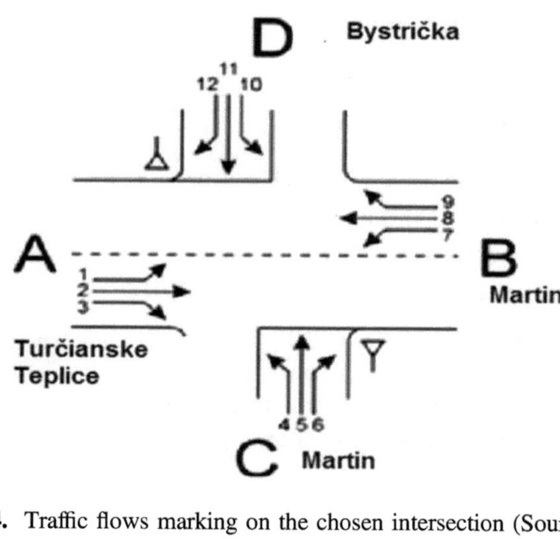

Fig. 4. Traffic flows marking on the chosen intersection (Source [2])

Traffic flows are divided into four levels depending on the duty to give way (Table 1):

- 1st level - main flows in forward direction from main direction, right turn from main direction,
- 2nd level - once subordinated flows left turn from main direction, right turn from side road,
- 3rd level - twice subordinated flows forward direction from side road,
- 4th level - triple subordinated flows left turn from side road.

Particular levels of quality mean:

- level A: Most of road users can pass the intersection without any affect. The delay time is very short,
- level B: Vehicles on the entrance to the intersection are partly affected. The delay time is short,
- level C: The delay time is longer but acceptable. Isolated short queues occur,
- level D: Significant time delays. State of traffic flow is stable, although temporary queues occur on entrances,
- level E: Congestion occurs and does not reduce within existing load. Time delays are very high. Sensitive dependence is characteristic. Small changes of a traffic load will trigger a rapid increase of time delays,
- level F: number of vehicles which enter intersection in time unit is bigger than capacity of intersection. Long queues of vehicles are made, time delays becoming unsustainable, intersection is overloaded.

Table 1. Evaluation of traffic flows quality.

Evaluation of traffic flows quality				
Traffic flow	Reserve of capacity R_i and R_m [pcu/h]	The average time delay w_i and w_m [s] or QSV		Comparison with the required time delay
1	526	< 10	A	Acceptable
7	486	< 10	A	Acceptable
6	403.5	< 10	A	Acceptable
12	442.5	< 10	A	Acceptable
5	1.2	> 45	E	Inappropriate
11	25.6	> 45	E	Inappropriate
4	16	> 45	E	Inappropriate
10	−70.9	–	F	Exceeded capacity
1 + (2 + 3)	995.5	< 10	A	Acceptable
7 + (8 + 9)	368.9	< 10	A	Acceptable
4 + 5+6	−89.3	–	F	Exceeded capacity
10 + 11 + 12	−162.8	–	F	Exceeded capacity

4 Modelling of the Chosen Intersection in the City of Martin

We have created a traffic model in the software tool Aimsun. Aimsun is traffic modelling software designed by Spanish company TSS - Transport Simulation Systems. Software allows making microscopic and macroscopic simulations. In Aimsun it is possible to simulate a variety of road networks: city networks, highways, expressways, bypass of the city and their combinations.

Software Aimsun generates several types of outputs during the whole simulation, which are divided into several groups: statistics of whole network, statistics of the chosen groups of road sections, statistics of an every section and an every turn, statistics of the chosen road, statistics of an origin/destination matrix, and statistics of public passenger transport. For an every chosen group the following results are generated: intensity, density, an average speed, delay times, a number of stops, stop times, speed and travel time, total distance and total travel time, a fuel consumption and an amount of produced emissions [3, 4].

4.1 Model Creation in Software Aimsun

The process of model creation consists of several steps. The first one was to import a base map. Next step was to define a road network, traffic flows and signal plans. Figure 5 shows the sequence of steps.

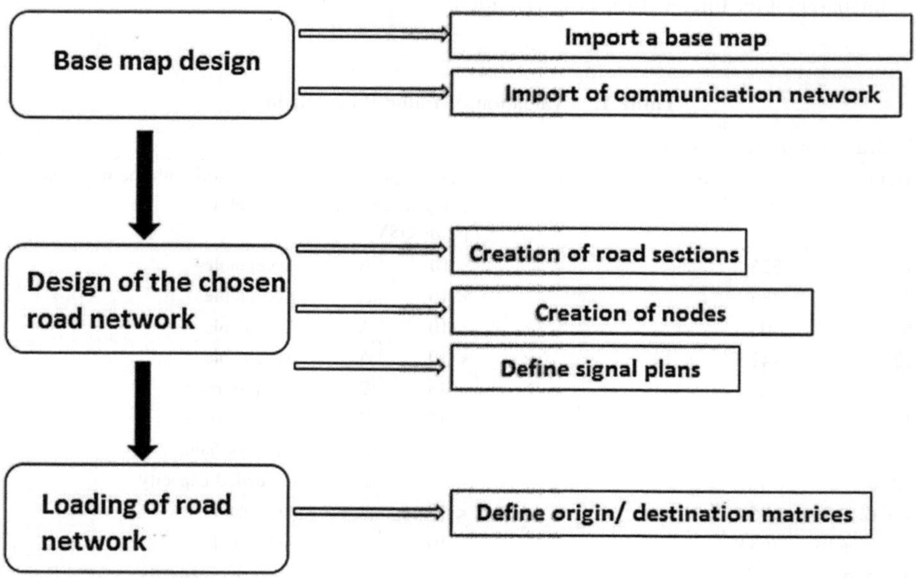

Fig. 5. The sequence of steps - creating model

Base map design - the fisrt step of base map design was to import a communication network of the chosen road network. We used OpenStreetMap due to the right scale. Then, pictures of the chosen part was imported.

Design of the chosen road network - using functions of Aimsun there were step by step created and edited each entrances of the intersection. We edited width of lanes, organising of lanes and also length of each sections and maximum speed [8].

4.2 Modelling Results

Based on the measured peak and saddle values, we made a simulation using the Aimsun. The results for entrance D (direction from Bystrička) are in the following tables (Table 2).

From the tables you can see:

- right turn from direction of Bystrička is the least problematic. The average delay in the morning peak hour is 113.73 s/vehicle and in the afternoon peak hour is 65.42 s/vehicle. Saddle hours are negligible,
- left turn from direction of Bystrička to Martin is very difficult. The average delay in the morning peak hour is 138.04 s/vehicle, even in some interval about 7:55 a.m. is delay 163.6 s/vehicle. This is very high value of delay, therefore it is necessary to solve this direction. In the afternoon peak hour the average delay is 84.8 s/vehicle. Saddle hours achieved the average delay 19.64 s/vehicle,
- pass through intersection forward from Bystrička has the average delay in the morning peak hour about 121.9 s/vehicle and in the afternoon peak hour it is 76.97 s/vehicle. Saddle hours achieved the average delay 18.96 s/vehicle.

On Fig. 6, there is visualization of traffic situation of left turn from the direction of Bystrička to Martin. You can see on this picture long queue of vehicles from the side direction.

Fig. 6. Visualization of vehicle queue on Entrance D in the morning peak hour

Table 2. The evaluation of traffic flows quality during the morning peak hour.

Entrance D - right turn (D - A)		
Morning peak hour 7:15–8:15		
Simulation time	Time delay [s]	Speed [km/h]
7:25:00	105.47	24.92
7:35:00	162.88	17.19
7:45:00	89.11	26.84
7:55:00	136.07	21.06
8:05:00	103.21	26.14
8:15:00	83.94	23.64
Weighted average	113.73	23.27
Entrance D - left turn (D - B)		
Simulation time	Time delay [s]	Speed [km/h]
7:25:00	79.02	26.54
7:35:00	160.71	18.29
7:45:00	147.07	21.09
7:55:00	163.60	19.84
8:05:00	154.14	21.92
8:15:00	105.06	21.53
Weighted average	138.04	21.25
Entrance D - forward (D - C)		
Simulation time	Time delay [s]	Speed [km/h]
7:25:00	92.10	23.81
7:35:00	176.33	16.48
7:45:00	93.07	24.38
7:55:00	161.37	19.47
8:05:00	104.93	24.26
8:15:00	103.82	21.23
Weighted average	121.90	21.63

5 The Possibilities to Increase Safety on the Chosen Intersection

We have designed two solutions to increase safety on the chosen intersection. The first one is to make a new horizontal marking shown on the Fig. 7. The second one is to change the organization on the intersection and build road traffic signalization with dynamic controlling. Both solutions were simulated in Aimsun. Calibration of traffic model was made based on the transport survey data and origin/destination matrices.

New marking - this intersection is marking very inappropriately and forces drivers to break traffic regulations. The engaging lane on the entrance B from direction of Martin is created and marking in wrong way. The combination of forward direction and left turn into the one lane is very inappropriate. On the Fig. 7, there is a current horizontal road marking in white color and in grey color, we design a new correct road

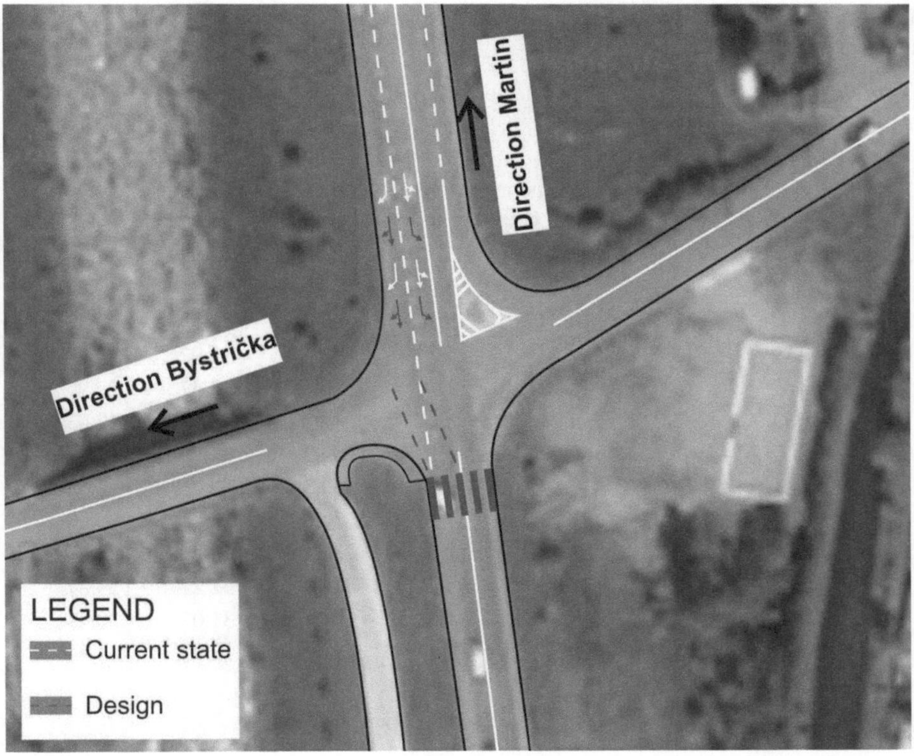

Fig. 7. Current horizontal road marking and design of new marking

markings. In this case would be appropriate to reduce speed to 50 km/h on the main road [6, 7]. These changes were implemented to the basic traffic model and some simulations were made.

These changes will solve the biggest problem of the intersection (entrance D) only partial, where time delays will be shorten especially in left turn from side direction (from Bystrička to Martin). Capacity evaluation shows that left turn needs to be solved immediately. Intensity of left turn exceeds the capacity of lane and achieves quality level F. The most suitable solution would be to change whole organization on the intersection, f. e. to build road traffic signalization with dynamic control. The aim of this measure is to increase safety of all road users, especially pedestrians. In the basic model, we have changed the uncontrolled junction to controlled junction with traffic signalization and made new simulations. The results of this simulated solutions are shown in the Table 3 [9].

Based on the comparison of each solutions (new marking and dynamic traffic controlling), we can say that it is possible to achieve savings over than 50% of time delay on entrance D. In case of building traffic signalization with dynamic control, it is possible to achieve savings, almost 80% of time delay on entrance D.

Table 3. Comparison of time delays - current state and possible solutions according to Aimsun simulations

Time delays [s]							
Current State				Change of lanes on entrance B		Built Traffic Light Signalization	
Entrance	Direction	Morning peak hour	Afternoon peak hour	Morning peak hour	Afternoon peak hour	Morning peak hour	Afternoon peak hour
A	A-B	3.69	2.97	3.40	2.85	15.89	12.49
	A-C	2.35	1.74	2.09	1.60	13.31	10.89
	A-D	7.86	10.79	5.76	9.09	21.53	22.40
B	B-A	12.02	7.79	5.23	3.44	14.92	12.49
	B-C	16.62	11.92	12.70	9.77	32.37	20.96
	B-D	5.97	3.74	5.86	3.97	14.51	12.03
C	C-A	60.41	60.23	43.17	39.33	43.50	38.87
	C-B	13.25	12.99	6.68	6.85	8.26	6.55
	C-D	43.57	37.18	24.90	22.96	35.47	34.19
D	D-A	113.73	65.42	51.13	28.91	28.91	29.44
	D-B	138.04	84.80	61.94	48.21	32.21	29.87
	D-C	121.90	76.97	63.08	43.26	31.07	30.48
Whole intersection		11.28	5.58	12.63	8.31	18.92	14.94

References

1. Kalašová, A., Faith, P., Paľo, J.: Dopravné Inžinierstvo. Žilinská Univerzita v Žiline. EDIS - Vydavateľstvo ŽU, Žilina (2006)
2. Ministry of Transport and Construction of the Slovak Republic: Technical Conditions TC 16/2015: Calculation of Road Communications Capacity. Ministry of Transport and Construction of the Slovak Republic, Bratislava (2015)
3. Poliak, M.: The Relationship with Reasonable Profit and Risk in Public Passenger Transport in the Slovakia. J. Econ. Theory Econ. Policy Social Econ. Forecast. **61**, 206–220 (2013)
4. Kalašová, A.: Štúdia Križovatky Ciest I/65 a III/ 2144, Obecný Úrad Bystrička. Study of Intersection of Roads I/65 and III/2144, Municipal Office. University of Žilina, Žilina (2016)
5. Website, http://www.mapy.cz
6. Gnap, J., Konečný, V.: The impact of a demographic trend on the demand for scheduled bus transport in the Slovak Republic. Communications **10**, 55–59 (2008)
7. Transport Simulation Systems: Aimsun 8. Dynamic Simulator User's Manual. Transport Simulation Systems (2017)
8. Smith, J., Blewitt, R.: Traffic Modelling Guidelines. Traffic Manager and Network Performance Best Practice. Version 3.0, http://www.tfl.gov.uk/assets/downloads/traffic-modelling-quidelines.pdf
9. Ondruš, J., Černický, Ľ.: Usage of Polcam device for parameter monitoring and traffic flow modelling. Communications **18**(2), 118–123 (2016)

Mixed Public Transport Lines Construction and Vehicle's Depots Location Problems

Piotr Sawicki(✉) and Szymon Fierek

Division of Transport Systems,
Poznan University of Technology, Poznań, Poland
{piotr.sawicki,szymon.fierek}@put.poznan.pl

Abstract. The paper deals with the problem of simultaneous consideration of transport lines construction and decision on vehicle's depots location problems in the public transport system. The authors have proposed a cost oriented mathematical model for solving a complex problem and application of macro-simulation based modified four-stage procedure for defining transportation tasks on the network. The proposed methodology is verified on a medium size testing transport network and conclusions have been finally formulated.

Keywords: Passenger transport · Public lines construction · Depots location · Optimization · Macroscopic traffic modelling · Visum

1 Introduction

1.1 Transport System Design

Public transport offers many advantages over individual ways of moving by private transport means, including: lower transportation costs, lower utilization of space per person travelled, lower energy consumption and lower pollution. Public transport increases overall mobility of the local community, especially of its part that does not possess private cars. It also improves accessibility to different destinations, including: places of employment, business activities centres, points of interests and/or recreational areas. For these reasons, local authorities especially in metropolitan areas are vividly interested in providing satisfactory level of public transport services, resulting in the overall increase of the urban life standard. Thus, designing of the mass transit system is crucial issue in urban management.

Public transport system design may be considered from different points of view depending of managing entity, i.e.:

- mass transit operator – operational management,
- public transportation board – tactical management,
- city authorities – strategic management.

On above mentioned levels appear such problems as crew scheduling, vehicle scheduling and line route planning with timetable design. Many authors confirm this statement, e.g. Ceder [1] and Schobel [2] claim that the most common is division of

E. Macioszek and G. Sierpiński (eds.), *Recent Advances in Traffic Engineering for Transport Networks and Systems*, Lecture Notes in Networks and Systems 21,
DOI 10.1007/978-3-319-64084-6_20

planning process in public transportation for several phases such as line route planning, timetabling, and vehicle and crew scheduling.

From methodological point of view different approaches are used to solve above mentioned problems. Lownes and Machemehl [3] proposed mixed integer model for a single-route circulator design problem, while Szeto and Wu [4] used metaheuristic algorithms (hybrid genetic algorithm) to solve simultaneously both the route design and frequency setting problems. Cancela et al. [5] propose to solve the problem of calculating the number and itinerary of bus routes and their frequencies for a public transportation system, using mathematical programming.

In this paper, the authors consider the line route planning allied with location of vehicle's depots. This approach is quite novel and has not been discussed in the literature yet. The problem of depots in the transport network is considered only through the deadheading context. Eberlein et al. [6] have considered a problem for deadheaded vehicles starting their trip empty until a station to be determined, and from that stop, these buses start their normal service until the end of their route.

1.2 Objective of the Paper

The paper is concentrated on the problem of simultaneous consideration of public transport lines construction and decision on vehicle's depots location problems. The authors defined a single criterion optimization model for simultaneous solving both considered problems. In the proposed methodology a modified macro-simulation based four-stage procedure for defining transport tasks on the network is involved [7]. This approach is verified on testing transport network composed of around 500 arcs and above 280 nodes.

The proposed methodology is an extension of a series of previous publications on solving vehicle's depot location problem (in fact tram depot location problem) with an application of combine optimization and multiple criteria decision aiding, e.g. [8–10]. All those works are performed with the principle assumption that a network of transport lines is given (redefined) and constant during problem solving.

2 The Proposed Methodology of Solving Combined Lines Construction and Depots Location Problems

2.1 General Assumptions

The essence of combining two strategic problems of the public transport system, i.e. lines construction and depots location, is to find a solution that guarantee the lowest operating costs for the operator, while keeping all the required standards for passengers. The nature of each decision problems is contradictory while are considered together. The problem of transport lines construction is strongly linked to the passengers' demand on movement, while depots location is dictated by searching the least expensive land costs, which means a location of depots on the suburbs. From the economical point of view it should be located as closed to each transport line as possible.

The combined solution of these two problems is primarily driven by the demand for mass transit and the location of the depots closest to the concentration points of a highly frequently utilized transport lines.

In the proposed methodology of simultaneous solving the combined lines construction and depots location problems, the authors have defined the following assumptions:

- lines construction is based on detailed analysis of trips performed on the considered area with rejection of any prior modal split,
- all the considered lines should be symmetric,
- a fleet size to handle each line is homogeneous, i.e. the capacity of each vehicle is equal,
- a capacity of the planned depots is unconstrained, i.e. the depot size results from optimal-based distribution of the fleet, due to the constructed route lines.

2.2 Principles of the Methodology

A General Scheme of the Methodology: The proposed methodology is composed of six consecutive steps (see Fig. 1). In the step 1, trips on the considered area are generated. The basis for this purpose is knowledge of the inhabitants' transport behaviour, typical for the analysed area. The result of step 1 is the amount of production and attraction established for each of the elementary parts of the considered area (so called traffic analysis zones). On that basis, in step 2, travel potentials are distributed between pairs of zones (origin and destination), and then a draft version of trips assignment to the individual arcs of the network is performed. Next, in the step 3 the lines structure is constructed (see step 3.1) in conjunction with a fleet sizing defined for each line (see step 3.2).

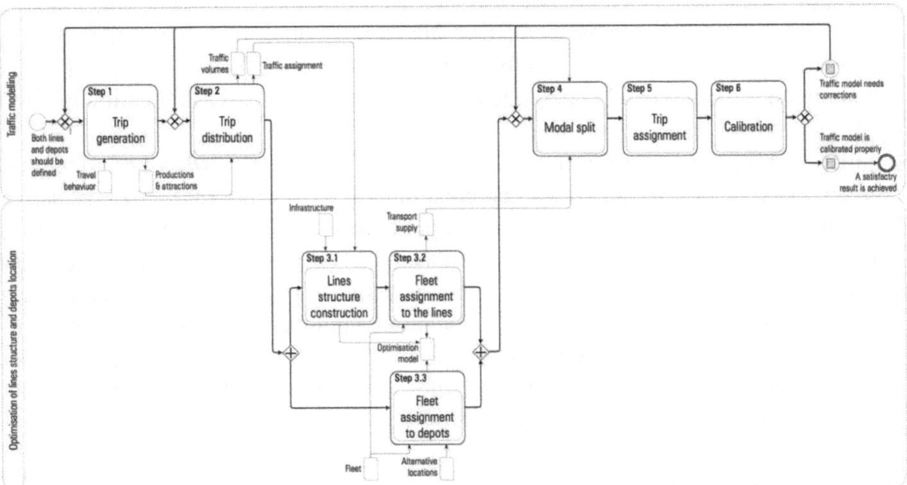

Fig. 1. The concept of the methodology of combined solving transport lines construction and depots location problems

At the same time, the fleet needed to perform a transport service is deployed to the most suitable depots. The basis for the implementation of steps 3.1–3.3 is the optimization model (see the following sub-sections).

As a result of the combined optimization of the transport lines construction and depots location (step 3), the volume of supply is defined. The supply volume in step 4 is the basis for determining the modal split, i.e. the division into private and public transport. In step 5, all the trips performed by inhabitants are assigned according to the defined modal split, and during step 6 transport model is calibrated. If the result of calibration is satisfactory, the calculation procedure is terminated, otherwise the previous steps (1 or 2 or 4) are performed again and the subsequent steps of the procedure are repeated until the result is satisfactory.

Realization of steps 1, 2 and 4 to 6 is performed using a dedicated macro-simulation tool for traffic modelling. Overview of these types of tools in the context of their key functionalities is presented in [11]. Step 3 (i.e. 3.1–3.3) is run in an optimization tool with a dedicated engine for linear programming problems.

Mathematical Model Overview: Since, in the defined methodology a simultaneous solving a complex decision problem (mixed transport lines and depot location problems) is assumed, in this section a construction of the mathematical model is suggested.

The objective function is formulated as an operational cost C of transport network operations, which is composed of both, transportation tasks performance costs TP and losses L with respect to the empty runs between depots and transportation lines.

Notations and Nomenclature: In the considered problem a following notations are assumed:

- indexes: i - transport line ($i = 1,\ldots, I$), j - a number of starting node on transportation network ($j = 1,\ldots, J$) and $j \neq k$, k - a number of end node on transport network ($k = 1,\ldots, K$) and $k \neq j$, l - location of a depot ($l = 1,\ldots, L$), m - vehicle to perform transport task ($m = 1,\ldots, M$), t - time interval ($t = 1,\ldots, T$),
- decision variables: y_{it} - a number of routes on i-th line performed during t interval [1/hour], $y_{\text{it}} \in \text{C}$; x_{ijkt} - an assignment of arc (j, k) to the i-th transport line during t interval [-]; $x_{\text{ijkt}} \in \{0,1\}$; z_{l} - allocation of l-th depot [-], $z_{\text{l}} \in \{0,1\}$; z_{ilm} - an assignment of m-th vehicle operating on i-th transport line to a l-th location of depot [-], $z_{\text{ilm}} \in \{0,1\}$,
- parameters: c_{ijk} - unit transport cost on arc (j, k) of i-th line [PLN/km], d_{jk} - length of arc (j, k) [km], q_{i} - capacity of a vehicle assigned to i-th line [pas.], λ_{i} - passenger comfort factor [-], q_{l} - capacity of l-th depot [veh.], P_{jkt} - passenger load on arc (j, k) in t-th time interval [pas./h], d_{il}^{p} - distance between l-th depot location and a first stop of i-th transport line [km], d_{il}^{k} - distance between l-th depot location and a last stop of i-th transport line [km], N - maximum number of depots in a transport system [-], r_{it} - number of vehicles assigned to i-th transport line during t-th period of time [veh./h], v_{it} - velocity of the vehicles at i-th transport line during t-th period of time, upon given traffic conditions [km/h].

Objective Function: The objective function in this problem is a minimized cost function, expressed as:

$$C = \min(TP + L) \tag{1}$$

where:

$$TP = \sum_{i=1}^{I}\sum_{j=1}^{J}\sum_{k=1}^{K} c_{ijk} \cdot d_{jk} \cdot \sum_{t=1}^{T} x_{ijkt} \cdot y_{it}, \ \text{[PLN/24 h]} \tag{2}$$

$$L = \sum_{l=1}^{L} z_l \sum_{i=1}^{I}\sum_{m=1}^{M} z_{ilm} \cdot \left(d_{il}^{p} + d_{il}^{k}\right), \ \text{[PLN/24 h]} \tag{3}$$

Constraints: The set of feasible solutions is constructed with the following formulations:

- a total capacity of the fleet of vehicles and its operational frequency are fundamental parameters for serving passenger volume on arc (j, k) of i-th transport line (see exp. 4),
- a total number of depots in transport system is limited to N (see exp. 5),
- all the vehicles in a transport system, i.e., assigned to all transport lines, have to be assigned strictly to available depots (see exp. 6) and a maximum number of vehicles available at i-th transport line is resulted from exp. (7).

$$\sum_{i=1}^{I} \lambda_i \cdot q_i \cdot x_{ijkt} \cdot y_{it} \geq P_{jkt}; \quad \forall j = 1,\ldots,J;\ k = 1,\ldots,K;\ j \neq k;\ t = 1,\ldots,T \tag{4}$$

$$\sum_{\mathrm{l}=1}^{\mathrm{L}} z_l \leq N \tag{5}$$

$$\sum_{i=1}^{I}\sum_{l=1}^{L}\sum_{m=1}^{M} z_{ilm} \leq \sum_{l=1}^{L} z_l \cdot q_l \tag{6}$$

$$\max_{1<t\leq T} r_{it} = \max_{1<t\leq T}\left[\left(y_{it}\sum_{j=1}^{J}\sum_{k=1}^{K} x_{ijkt} \cdot 2d_{jk}\right)\Big/ v_{it}\right], \ \forall\ i = 1,\ldots,I \tag{7}$$

3 Verification of the Methodology

3.1 Computational Experiments

A verification of the methodology of combined public transport lines construction and vehicle's depots location problems is performed on the basis of testing transport system (TTS) of a medium sized city in Poland. In the computational experiments, the authors have utilized transport demand modelling software - PTV Visum and optimisation engine for MS Excel - Solver.

In the step 1 (Trip generation) several actions have been carried out. One group is focused on transport infrastructure definition (see Fig. 2.), i.e. all the nodes are enumerated (total number of nodes is 210), active arcs (*j*, *k*) are identified (381 arcs) and length of each arc d_{jk} is measured (see the result in Table 1). The second group of actions is focused on trip description, i.e. traffic zones are extracted and their travel potential (production and attraction for each zone) are defined. It is based on economic and demographic characteristics of each traffic zone.

Step 2 (Trip distribution) is featured by determination of all trips between any pair of origin and destination zones on the considered area. The result is a complete overall O-D matrix and a draft of traffic assignment on the network P_{jkt} is performed. A sample result of the draft trip distribution is presented in Table 1.

Table 1. A sample of draft trip distribution

Nodes		Volume [pas./24 h]		d_{jk} [km]	v_{jk} [km/h]	Nodes		Volume [pas./24 h]		d_{jk} [km]	v_{jk} [km/h]
j	k	P_{jkt}	P_{kjt}			j	k	P_{jkt}	P_{kjt}		
1	2	95	35	.11	30	140	133	1143	720	1.61	11
1	6	811	889	.80	17	140	139	2868	4439	3.31	7
1	115	816	799	.58	17	140	142	2907	1662	.98	7
2	6	851	931	.74	17	141	133	480	645	.80	21
3	62	1620	1675	.87	11	141	138	1119	1021	.40	17
…	…	…	…	…	…	…	…	…	…	…	…
70	71	614	600	.20	18	207	52	1638	1731	.32	11
70	116	913	894	.17	16	207	60	639	639	.11	18
70	200	280	313	.40	26	208	207	999	1091	.06	17
71	67	473	400	.30	22	209	204	891	713	.17	17
71	197	141	200	.35	29	209	208	999	1091	.17	17
72	2	42	173	1.02	52	209	210	201	286	.20	25
72	5	2207	2260	.24	8	210	205	536	333	.15	20
72	206	2242	2057	.28	8	210	206	73	360	.06	48

During step 3 an optimisation model has been applied and parameterised first. The complete list of parameter values is presented in Table 2.

Table 2. A set of parameters of optimization model

Parameter			Comments
Notation	Unit	Value	
c_{ijk}	[PLN/km]	5.2	Unit cost in medium sized transport operator
q_i	[pas.]	105	A total capacity of exemplary city bus - Solaris Urbino12
λ_i	[-]	.75	A passenger comfort factor, i.e. at most 75% of available capacity should be utilised
N	[-]	1	A limited number of depots in transportation system

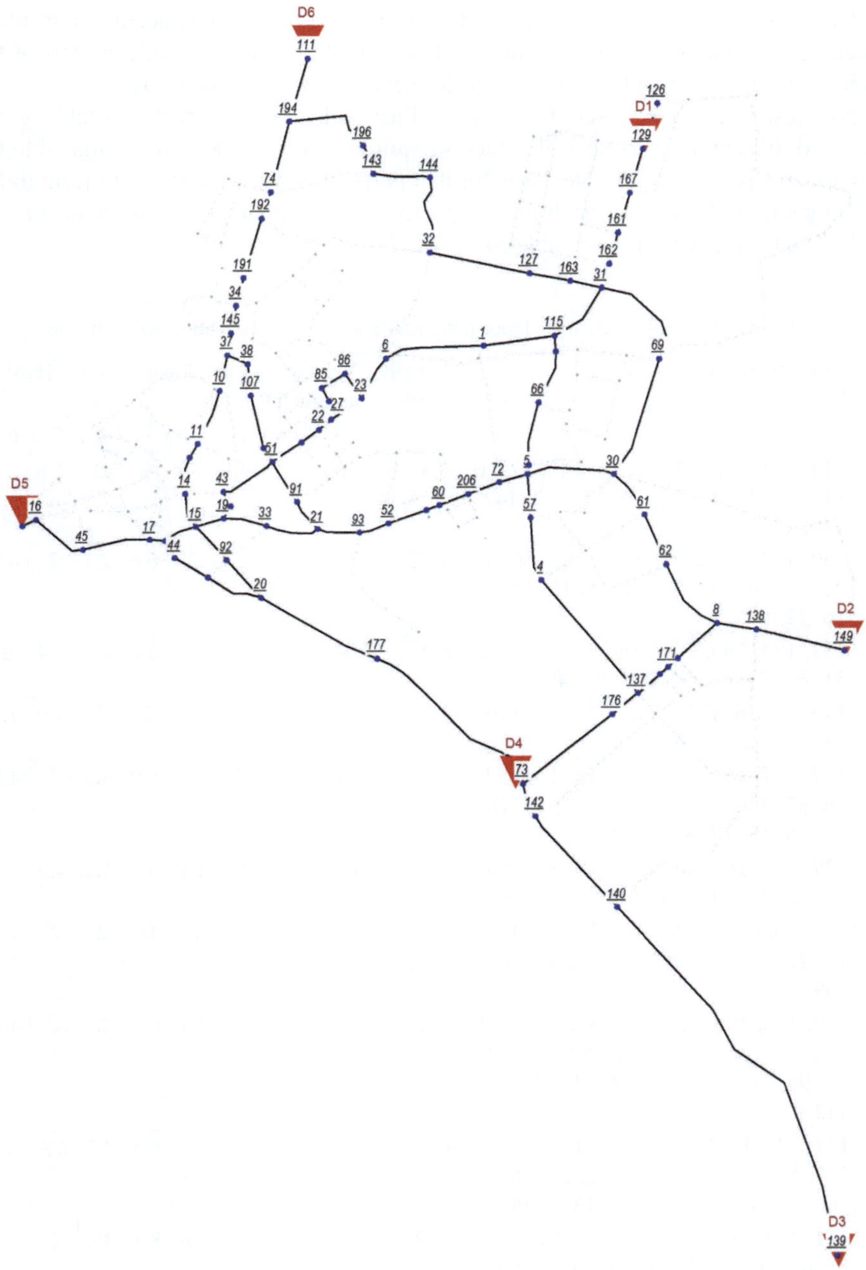

Fig. 2. Simplification of considered transportation network with possible localization of depots (red triangle: D1–D6). More detail network is in the background (grey links and nodes)

During step 3.1, a structure of the transport lines has been constructed as a result of searching for minimal number of lines that are able to handle a significant part of the traffic assigned to the individual arcs of the network, at the lowest cost.

As a result in Table 3 a set of 10 selected lines and their one-direction total length is presented. In step 3.2 a number of vehicles required to handle the traffic assigned to the network arcs is determined. The basis for that purpose is: velocity resulting from traffic conditions (see Table 1 for each arc), capacity of the vehicle and comfort factor (see Table 2) and line length (see Table 3).

Table 3. A set of lines, line-depot distances and vehicle allocation to each line

No	Lines (node sequence)	Length [km]	Vehicles assigned [veh.]	Line-depot distance [km] for *l*:=					
				1	2	3	**4**	5	6
1	**111**, 194, 74, 192,191, 34, 145, 37, 10, 11, 118, 14, 15, 92, 20, 177, 73, 142, 140, **139**	6.3	26	4.0	3.1	.2	**.2**	1.8	.2
2	**149**, 138, 8, 62, 61, 30, 5, 72, 206, 60, 207, 52, 93, 21, 33, 19, 15, 18, 17, 45, 16, **122**	10.2	7	2.9	.2	6.6	**2.1**	.2	4.0
3	**111**, 194, 196, 143, 144, 32, 127, 163, 31, 69, 30, 61, 62, 8, 138, **149**	8.0	14	1.2	.2	6.6	**2.1**	5.4	.2
4	**149**, 138, 8, 172, 171, 170, 137, 176, 73, 142, 140, **139**	2.7	6	4.4	.2	.2	**.2**	4.9	7.4
5	**129**, 167, 161, 162, 31, 115, 1, 6, 23, 86, 85, 88, 87, 26, 22, 24, 51, 43, 181, 19, 15, 18, 17, 45, 16, **112**	8.4	25	.2	4.4	8.0	**3.5**	.2	4.0
6	**129**, 167, 161, 162, 31, 115, 116, 66, 56, 5, 57, 4, 137, 134, **133**	6.0	17	.2	1.9	5.7	**1.2**	4.6	4.0
7	**112**, 16, 45, 17, 18, 44, 101, 20, 177, 73, 176, 137, 170, 171, 172, 8, 138, **149**	8.1	21	4.4	.2	4.6	**.2**	.2	4.3
8	**139**, 140, 142, 73, 176, 137, 170, 171, 172, 8, 62, 61, 30, 5, 72, 206, 60, 207, 52, 93, 21, 33, 19, 15, 18, 17, 45, 16, **112**	13.3	8	2.9	1.1	.2	**.2**	.2	4.0
9	**111**, 194, 74, 192, 191, 34, 145, 37, 38, 107, 54, 51, 91, 21, 93, 52, 207, 60, 206, 72, 5, 30, 61, 62, 8, 138, **149**	11.0	34	4.0	.2	6.6	**2.1**	2.9	.2
10	**111**, 194, 196, 143, 144, 32, 127, 163, 31, 69, 30, 5, 72, 206, 60, 207, 52, 93, 21, 33, 19, 15, 18, 17, 45, 16, **112**	12.7	24	.2	2.6	8.1	**3.6**	.2	.2

The resulted total number of 182 vehicles distributed to 10 lines is shown in Table 3. Finally, during step 3.3 all the vehicles resulted from step 3.2 are allocated to one of the potential depot ($N = 1$). Based on the global optimisation model (step 3.1-3.3 are linked each other and computation is performed simultaneously) a minimum cost objective function has been achieved: $C = 20{,}630.0$ PLN, with $TP = 17{,}426.3$ PLN and $L = 3{,}203.7$ PLN, respectively. As a result a D4 (see Fig. 2) vehicle's depot is selected ($l = 4$) and all 182 vehicles should be served by that location.

In the step 4 (modal split) calculation of shares of trips made by public as well as private transportation means was performed. These computations were made based on

Table 4. A traffic allocation to arcs of the network for morning rush hours (7 am–8 am) for public transport only

Nodes		Volume [pas./h]		d_{jk} [km]	v_{jk} [km/h]	Nodes		Volume [pas./h]		d_{jk} [km]	v_{jk} [km/h]
j	k	P_{jkt}	P_{kjt}			j	k	P_{jkt}	P_{kjt}		
1	2	0	0	.11	35	140	133	47	0	1.61	35
1	6	1	47	.80	35	140	139	371	573	3.31	43
1	115	47	1	.58	45	140	142	475	475	.98	47
2	6	0	0	.74	35	141	133	0	47	.80	35
3	62	77	84	.87	44	141	138	109	100	.40	31
…	…	…	…	…	…	…	…	…	…	…	…
70	71	0	0	.20	45	207	52	411		.32	32
70	116	0	0	.17	45	207	60	287		.11	37
70	200	0	0	.40	45	208	207	135	167	.06	29
71	67	0	0	.30	45	209	204	167	135	.17	27
71	197	0	0	.35	35	209	208	135	167	.17	29
72	2	0	0	1.02	35	209	210	0	0	.20	35
72	5	329	390	.24	32	210	205	17	0	.15	35
72	206	390	329	.28	32	210	206	0	17	.06	35

Table 5. Line structure, line-depot distances, vehicle allocation to each line and to the depot

No	Lines (node sequence)	Length [km]	Vehicles assigned [veh.]	Line-depot distance [km] for l:=					
				1	2	3	4	5	6
1	**111**, 194, …, 140, **139**	6.3	6	4.0	**3.1**	.2	.2	1.8	.2
3	**111**, 194, …, 138, **149**	8.0	2	1.2	**.2**	6.6	2.1	5.4	.2
4	**149**, 138, …, 140, **139**	2.7	1	4.4	**.2**	.2	.2	4.9	7.4
5	**129**, 167, …, 16, **112**	8.4	3	.2	**4.4**	8.0	3.5	.2	4.0
6	**129**, 167, …, 134, **133**	6.0	3	.2	**1.9**	5.7	1.2	4.6	4.0
7	**112**, 16, …, 138, **149**	8.1	2	4.4	**.2**	4.6	.2	.2	4.3
8	**139**, 140, …, 16, **112**	13.3	1	2.9	**1.1**	.2	.2	.2	4.0
9	**111**, 194, …, 138, **149**	11.0	5	4.0	**.2**	6.6	2.1	2.9	.2
10	**111**, 194, …, 16, **112**	12.7	3	.2	**2.6**	8.1	3.6	.2	.2

the skim matrix consists of journey time (between particular zones). Thus the transportation supply (transportation network, public transportation line routes with timetables etc.) and traffic conditions as well are crucial components. The result of this step is a set of O-D matrices separate for each transportation mode. In the considered situation, the ratio of trips made by public and private transportation is 42% to 58% respectively.

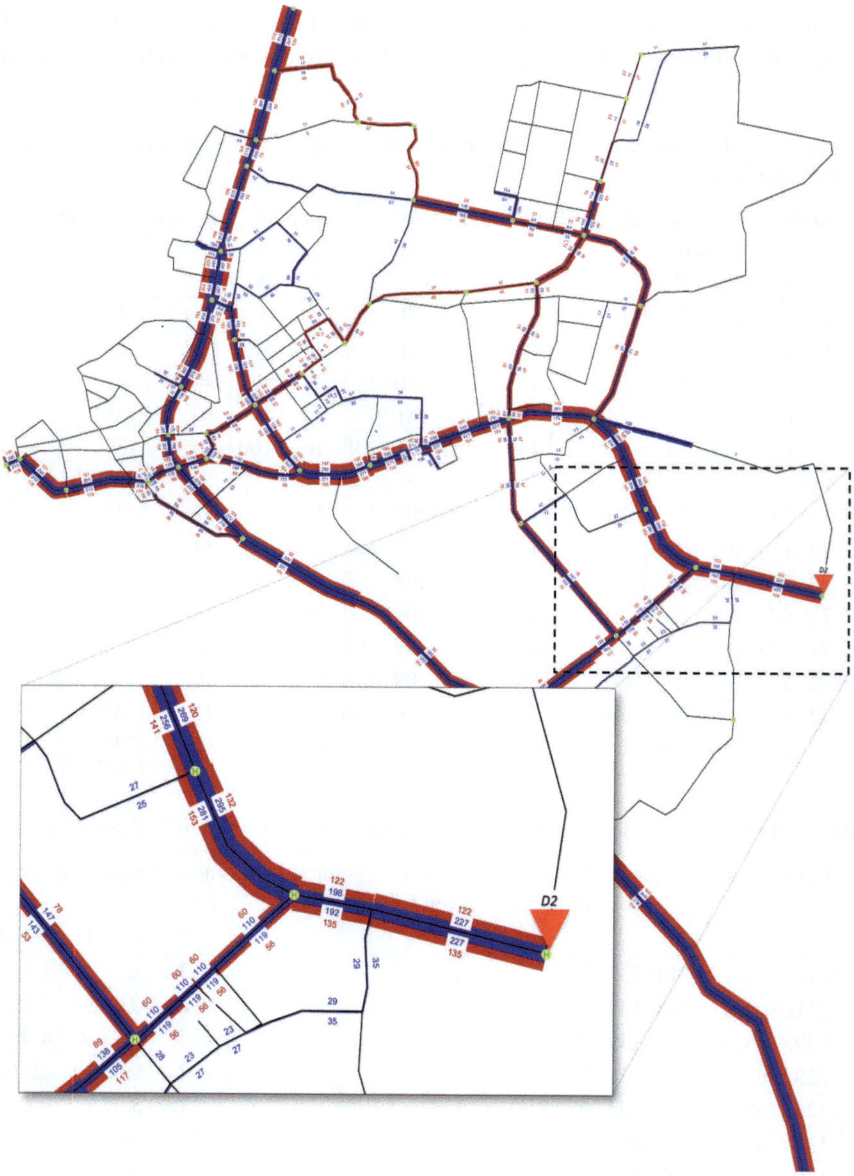

Fig. 3. A sample of final results - traffic volumes (blue bars and values – private transportation flows, red bars and values – public transportation flows)

Finally, in the last step 5 (traffic assignment), allocation of demand to supply has been carried out.

Since the first iteration assumed a traffic assignment in step 2 without modal split structure, further iterations are necessary to determine the final line construction and base location. As a result in Table 4 traffic allocation to the arcs of a considered network with modal split from the first iteration is presented. In fact traffic allocation during morning rush hour (7am–8am) is presented; it is 12.9% of 24-hour traffic.

During the second iteration, as a result of application the optimization model, the network structure is constructed with 9 (see Table 5) lines, i.e. line no 2 available in Table 3 is rejected here.

The fleet is constructed with 14 vehicles (see Table 5) and it is minimum number to provide line's operation during morning rush hour (7 am to 8 pm). For such a network, the best depot location is D2 ($l = 2$). The total operating cost for the morning rush hour plus cost of reaching the line and returning to the depot in this case is 2,824.3 PLN (TP = 2,567.3 PLN and L = 257.0 PLN).

Based on this result it is allowed to interpret a final results i.e. the volumes of passengers' and vehicles' flows on particular arc of the considered network (see Fig. 3).

4 Conclusions

4.1 Discussion on Results

The problem discussed in the paper consists in simultaneous optimization of the transport line system and the location of the depot in the urban transport system. The methodology proposed by the authors utilise a synergy of macroscopic traffic modelling and optimization methods. As a result of the iterative application of the methodology, a traffic allocation to the network determines both the structure of the lines and the location of the depot to handle all the utilised transport means, at the minimum global operating cost.

Since the traffic allocation to the network is resulted from real mobility pattern of the inhabitants, the final solution is characterized by a high level of supply matching (line configuration and its maintenance) to the real demand (mobility and inhabitants' behaviour).

4.2 Further Steps

At the present stage of the research three potential directions of a development of the proposed methodology can be formulated. First, it is proposed to reflect at the step 1 expected changes of transport behaviour in the long-term horizon, e.g. 5, 10 and 15 years perspective. The obtained results should allow determining the extent of the necessary changes in both considered problems, i.e. the lines structure and the depots location, taking into account predicted changes in mobility.

Another direction of the research is the extension of the scope of both combined problems with regard to the fleet composition problem. It means different capacity of transport means and sizing of each type should be additionally taking into account. This problem has been extensively analysed by one of the author in his previous research. Incorporating fleet composition should allow a better matching of supply to demand, and could directly result in better utilization of available transport capacity while reducing fleet operating costs at the same time.

The last potential direction of development concerns the application of the methodology into different types and real-life transport systems, including tram and bus transport. As a result the universality of the methodology and the measurable benefits of its practical application should be validated. The application of the proposed methodology into the real-life transport network should also result in carrying out step 6 of the methodology, i.e. calibration. In the problem considered in the paper a real traffic condition is not available, thus a calibration of the traffic model is inapplicable.

References

1. Ceder, A.: Public Transit Planning and Operation: Theory, Modelling And Practice. Elsevier, Oxford (2007)
2. Schobel, A.: Line planning in public transportation: models and methods. OR Spectr. **34**, 491–510 (2012)
3. Lownes, N., Machemehl, R.B.: Exact and heuristic methods for public transit circulator design. Transp. Res. Part B **44**(2), 309–318 (2010)
4. Szeto, W.Y., Wu, Y.: A simultaneous bus route design and frequency setting problem for Tin Shui Wai, Hong Kong. Eur. J. Oper. Res. **209**(2), 141–155 (2011)
5. Cancela, H., Mauttone, A., Urquhart, M.E.: Mathematical programming formulations for transit network design. Transp. Res. Part B **77**, 17–37 (2015)
6. Eberlein, X.J., Wilson, N.H.M., Barnhart, C., Bernstein, D.: The real-time deadheading problem in transit operations control. Transp. Res. Part B **32**(2), 77–100 (1998)
7. Fierek, S., Żak, J.: Planning of an integrated urban transportation system based on macro-simulation and MCDM/A methods. Procedia **54**, 567–579 (2012)
8. Kupka, P., Sawicki, P.: Optymalizacja Lokalizacji Zajezdni Tramwajowej w Systemie Komunikacji Miejskiej. Logistyka **2**, 462–472 (2015)
9. Sawicki, P., Kupka, P.: Ustalenie Lokalizacji Zajezdni Tramwajowej w Systemie Komunikacji Miejskiej z Zastosowaniem Metody Wspomagania Decyzji. Urban Regional Transport **6**, 6–11 (2016)
10. Sawicki, P., Kiciński, M., Fierek, S.: Selection of the most adequate trip-modelling tool for an integrated transport planning system. Arch. Transp. **37**(1), 55–66 (2016)
11. Redmer, A., Żak, J., Sawicki, P., Maciejewski, M.: Heuristic approach to fleet composition problem. Procedia **54**(4), 414–427 (2012)

Computer Support of Decision-Making for the Planning Movement of Freight Wagons on the Rail Network

Marianna Jacyna[1(✉)] and Mirosław Krześniak[2]

[1] Faculty of Transport, Warsaw University of Technology, Warsaw, Poland
maja@wt.pw.edu.pl

[2] PKP CARGO S.A. Poland, Warsaw, Poland
miroslawkrzesniak@gmail.com

Abstract. The article presents selected aspects of the problem of supporting the planning of freight wagons on the rail network. It has been pointed out that one of the problems in freight transport is the proper planning of operational activities of rail carriers. The problem is difficult because of its complexity. Planning the movement of freight wagons should take into account the need not only to transport loaded wagons, but also empty ones. The complexity of the problem stems from a number of factors, which should be considered during transport planning. These factors can be divided into technical, organizational or economic. The analysis allowed to develop a mathematical model for planning the movement of loaded and empty wagons, which includes: data schemes, decision variables, boundary conditions and constraints, criterion functions, as well as traffic disruptions in the rail network. Consequently, it was possible to develop a heuristic algorithm and a simulation tool, in the form of the ModPCar application. The subject matter is connected mainly with business needs of railway carriers. The effect, in the form of an IT tool, facilitating the process of decision support in the organization of rail transport, may be used, among others, by rail freight carriers. This paper describes the stages of the ant algorithm i.e. the stage of designating the probability of transition of ants to the other points of the route, update pheromone. In this paper the process of calibration of this algorithm was presented. The results of the ant algorithm were compared with the random results.

Keywords: Railway transport planning · Compact system · Railway operational service · Wagons rotation

1 Introduction

Planning for the displacement of full and empty wagons, being one of the elements of railway freight organization, requires prior assessment of transport requirements. Hence, transportation tasks are closely related to the transport needs of customers, which for a given period can be expressed in the amount of freight intended for transport between particular railway stations, or in the number of wagons needed to meet the transport demand. Determining the volume of transport tasks (transport needs) is necessary to develop freight traffic plans. Demand for transportation reported by customers is the transport system feed stream and is reported by the surrounding environment [1, 2].

E. Macioszek and G. Sierpiński (eds.), *Recent Advances in Traffic Engineering for Transport Networks and Systems*, Lecture Notes in Networks and Systems 21,
DOI 10.1007/978-3-319-64084-6_21

The process of planning the movement of full and empty wagons is a complex decision problem. First of all, it is necessary to take into account the efficient realization of the set objectives, meaning the fulfillment of requested transport needs. The decision-making complexity of planning is a result of the multitude of problems and a number of factors, which should be considered. These factors can be divided into technical, organizational or economic. Compact system of transport is connected with different problems than dispersed transport. The transportation technology used depends on the size of the batch of goods requested by customers to be transported. For large batches, transport is carried out in a compact system, and for small batches a dispersed system is rather used.

Compact system is the most cost-efficient system for freight. The identifying feature of this system is that the entire group of wagons is moved as one train, essentially using one motor unit, without the need for costly shunting at intermediate stations. In Fig. 1 red color indicates the process of compact transportation.

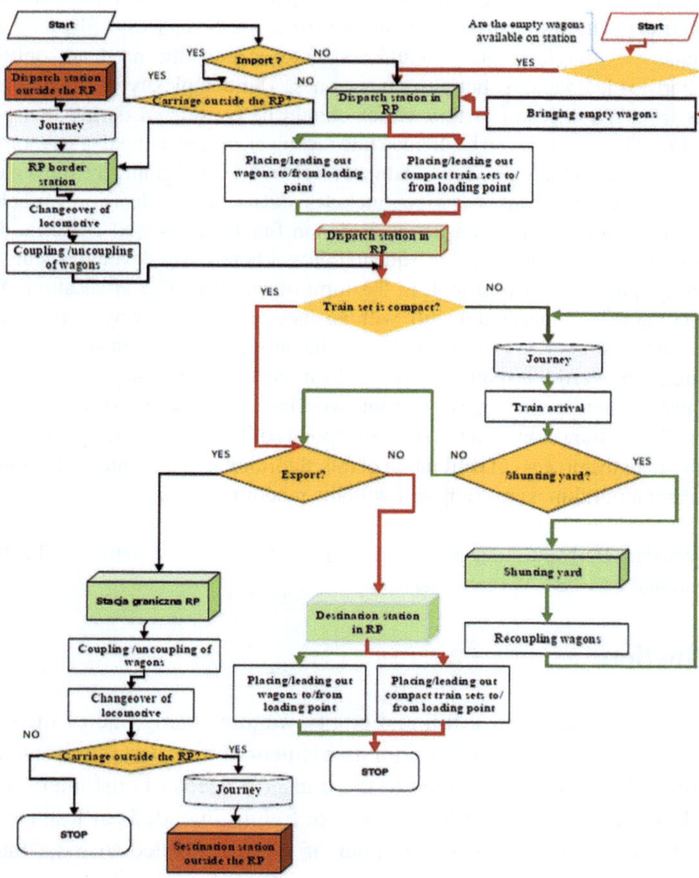

Fig. 1. Technology of moving wagons on the railway network

Such a system of transport increases its quality and speeds up the delivery. This means that it is the most efficient form of freight, and thus is most desirable from the point of view of the carrier.

Another way of moving wagons is a dispersed system. The transportation of dispersed goods is most often carried out according to the following scheme: sending station - shunting station in the area of the sending station - marshalling station in the area of the sending station - marshalling station in the area of the destination station - shunting station in the area of the destination station - the destination station. Shunting staff and locomotives for delivering and picking up wagons between loading points in the station's area are concentrated on each shunting station. In Fig. 1 green color indicates the process of dispersed transport.

Dispersed shipments may also be transported in the system of linear trains. This type of transport technology assumes launching a train designed to collect cargo in one relation, when a compact train cannot be started due to insufficient volume of goods. The organization of this type of transport is based on a study of the potential demand for transport and running regular train connections according to a published timetable. Customers provide loads to stations along the way of the train, according to the pre-arranged timetable. Assuming that the starting station is also the end station of a given train connection, it is possible to talk about trains running in closed logistic loops. Adopting such an organization of transport of dispersed batches of goods allows to designate the rolling stock for this operation.

2 Research Matter in Literature

Dispersed shipments may also be transported in the system of linear trains. This type of transport technology assumes launching a train designed to collect cargo in one relation, when a compact train cannot be started due to insufficient volume of goods.

Legal regulations in the field of rail transport in Poland impact the behavior of entities such as: Railway Transport Authority (UTK), infrastructure managers (PKP Polish Railway Lines) and railway carriers. The most important act is the Railway Transport Act [3], which defines the rights and obligations of the listed market participants. Traffic on the railway network is determined by various conditions. In order to make rail transport work properly, the Railway Transport Act [3] identifies the activities that make up the management of rail infrastructure.

The essential element of organizing railway traffic is to plan the timetable [4–7]. The train scheduling process can be divided into two stages. The first is the preparation of transport offer and the second - creating the timetable itself. Preparation of the transport offer requires extensive consultation between infrastructure manager and the carriers. Furthermore, the carriers are consulting their needs with the customers. The aim of each railway carrier is to meet customer expectations at the lowest possible cost. A decision problem arises of how to organize your work to achieve the goals which were set. Due to the complexity of the processes of transporting full and empty wagons in the transport network, especially the strong dependency between their location and a point in time, a simulation model was used to solve the problem. The model of the system [8, 9] is a quantitative and qualitative representation of the system under consideration, capturing those characteristics that are relevant to the research being conducted. The construction of the model requires a number of assumptions concerning, among others, the railway infrastructure, its characteristics and the relationships that describe the transport process.

In the literature, a lot of research regarding modelling transport systems can be found: [1, 8]. The problems of modeling rail transport systems can be divided into the following areas [8–13]:

- analysis and evaluation of the technology of operation used in the marshalling station and the holding station, as well as investigating the fluency of train movement and organizing traffic on the railway network,
- analysis and evaluation of cargo and passenger handling in the context of infrastructure being used,
- study of the dependence of the efficiency of transport processes on the parameters of line and point infrastructure, analysis of the capacity of infrastructure elements, analysis of timetables, etc.

When building a simulation model, it is important to remember that this is the end result of a series of activities of identifying and formalizing the characteristics of the real system, and of determining the relationships between these features over time to find their counterparts in the system model [10]. A crucial element in creating a simulation model is to reflect the relationships, dependencies and behavior of objects and their characteristics from the real system into the behavior of the modelled system. To solve the problem, it is necessary to use appropriate methods (algorithms), which can be divided into traditional algorithms and meta-heuristic algorithms.

Thanks to continuous advancements of IT solutions, numerical methods are currently often used to solve problems of high complexity or lacking available analytical methods [1, 11]. Linear and nonlinear problems are most commonly solved using these methods. In addition, multi-variant optimization methods are becoming increasingly popular [2, 9, 11, 14–18]. These methods use multiple sub-criteria when establishing the optimal solution.

Another group of optimization methods are heuristic methods [19–23], basing on expert approach to the problem, where experience in solving the optimization task is important. Analysis of existing literature proves that the research problem of optimizing the planning of the transport of full and empty wagons cannot be solved by classical numerical methods due to its complex character. Therefore, it was decided to use approximate heuristic algorithms to solve the problem.

3 The Model Movement of Full and Empty Wagons

3.1 Assumptions

Starting from the assumption that empty wagons should be moved in such a way that they arrive at the destination station, being the station where they are needed, in the shortest possible time and with the lowest cost, a suitable model for this problem should be formulated. The approach to modeling the organization of relocating full and empty wagons on the rail network presented in this paper is based on the approach to modeling transport systems presented in [1, 2]. The research problem concerns the organization of the movement of full and empty wagons.

Criteria for the evaluation of the movement of empty and full wagons - the criterion of the cost of executing the relocation plan was selected to assess the quality of the solution of the movement plan for empty and full wagons.

Based on the data held by the carriers, it is assumed that the quantities and types of wagons on the entire railway network are known, and whether they are loaded or empty in particular areas at the moment *t*, $t \in \boldsymbol{T}$. The main assumptions for the mathematical model was presented in the form of algorithm, Fig. 2.

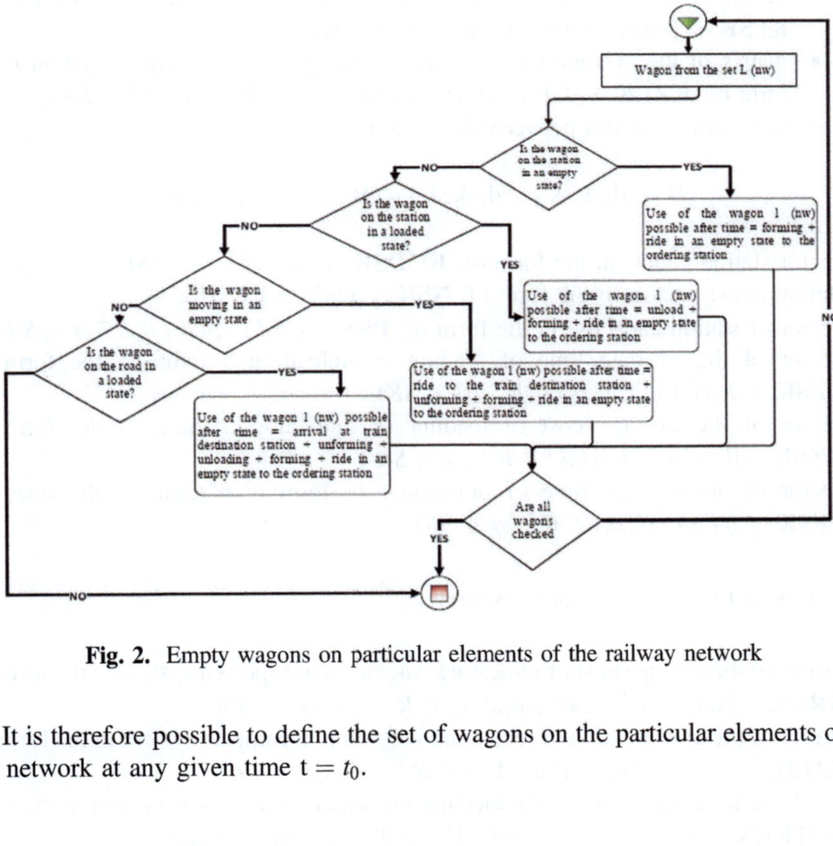

Fig. 2. Empty wagons on particular elements of the railway network

It is therefore possible to define the set of wagons on the particular elements of the rail network at any given time $t = t_0$.

3.2 The Input Data

The example data input essential to describe the mathematical model was presented below:

(a) railway stations:

- the set of $\boldsymbol{Z}$ numbers of transport orders will be a set of elements: $Z = \{1, \ldots, z, \ldots Z\}$,

- the set of $\boldsymbol{NW}$ wagon series numbers is a set in the form of: $\boldsymbol{NW} = \{1,\ldots, nw,\ldots NW\}$,
- the set of $\boldsymbol{SK}$ number of forwarding station and receiving station of wagons loaded and empty: $\mathbf{SK} = \{1,\ldots, \mathrm{sk},\ldots SK\}$,
- the set of $\boldsymbol{SN}$ number of forwarding station: $\mathbf{SN} = \{\mathrm{sk} \equiv \mathrm{sn}: \alpha(\mathrm{sk}) = 0 : \mathrm{sk} \in \mathbf{SK}\}$ $\alpha(\mathrm{v}) = \{0\}$ - the mapping which assigns the number of elements of the railway network to object type,
- the set of the $\boldsymbol{T}$ number of equal time periods (e.g. hours) $T = \{1,\ldots, t,\ldots T\}$,
- the set of the numer $\boldsymbol{L(nw)}$ of wagons nw-th series,
- the set of $\boldsymbol{LK}$ number of sections of the railway line, β - the mapping in set $\{0,1\}$: $\mathbf{LK} = \{(\mathrm{sk},\mathrm{sk}') : \beta(\mathrm{sk},\mathrm{sk}') = 1 : \mathrm{sk}, \mathrm{sk}' \in \mathbf{SK} \wedge \mathrm{sk} \neq \mathrm{sk}'\}$
- vector of the average cost of unforming or forming a train, in the form of $\mathbf{KFSK} = [\mathrm{kf}(\mathrm{sk}) : \mathrm{kf}(\mathrm{sk}) \in \mathbf{R}^+, \mathrm{sk} \in \mathbf{SK}]$,
- matrix of the average cost of relocating wagons in a compact system in the form of: $\mathbf{KZLK} = [kz((sk, sk')) : kz((sk, sk')) \in \boldsymbol{R}^+, (sk, sk') \in \boldsymbol{LK}]$,
- matrix of distances between the stations:

$$D = [\mathrm{d}(\mathrm{sk},\mathrm{sk}') : \mathrm{d}(\mathrm{sk},\mathrm{sk}') \in \mathbf{R}^+, (\mathrm{sk},\mathrm{sk}') \in \mathbf{LK}],$$

- vector station codes, in the form of: $\mathbf{KODSK} = [kod(sk) : sk \in \boldsymbol{SK}]$,
- station name vector, in the form of: $\mathbf{NSK} = [n(sk) : sk \in \boldsymbol{SK}]$,
- vector of station capacity, in the form of: $\mathbf{PSK} = [p(sk) : p(sk) \in \boldsymbol{C}^+ sk \in \boldsymbol{SK}]$,
- vector of the average time of loading or unloading a train, in the form of: $\mathbf{TZSK} = [tzw(sk) : tzw(sk) \in \boldsymbol{R}^+, sk \in \boldsymbol{SK}]$,
- vector of the average cost of loading or unloading a train, in the form of: $\mathbf{KZSK} = [kzw(sk) : kzw(sk) \in \boldsymbol{R}^+, sk \in \boldsymbol{SK}]$,
- vector of the average time of unforming or forming a train, in the form of: $\mathbf{TFSK} = [tf(sk) : tf(sk) \in \boldsymbol{R}^+, sk \in \boldsymbol{SK}]$,

(b) costs and times of relocating wagons:

- matrix of the average cost of relocating wagons in a dispersed system in the form of: $\mathbf{KRLK} = [kr((sk, sk')) : kr((sk, sk')) \in \boldsymbol{R}^+, (sk, sk') \in \boldsymbol{LK}]$,
- matrix of the average time of relocating wagons in a compact system in the form of: $\mathbf{TZLK} = [tz((sk, sk')) : tz((sk, sk')) \in \boldsymbol{R}^+, (sk, sk') \in \boldsymbol{LK}]$,
- matrix of the average time of relocating wagons in a dispersed system in the form of: $\mathbf{TRLK} = [tr((sk, sk')) : tr((sk, sk')) \in \boldsymbol{R}^+, (sk, sk') \in \boldsymbol{LK}]$,

(c). transport orders and wagon series:

- vector of the requested wagon providing dates in the form of: $\mathbf{TPZ} = [tpz(z, nw) : tpz(z, nw) \in \boldsymbol{R}^+, z \in \mathbf{Z}, nw \in \boldsymbol{NW}]$,
- vector of the requested sending dates of full wagons, in the form of: $TNZ = [\mathrm{tnz}(\mathrm{z},\mathrm{nw}) : \mathrm{tnz}(\mathrm{z},\mathrm{nw}) \in \mathbf{R}^+, \mathrm{z} \in \mathbf{Z}, \mathrm{nw} \in \mathbf{NW}]$.

3.3 Decision Variables

In case of the problem covered by this research, the decision variable is a number of wagons that will be used to complete the order. Therefore, the developed model includes 4 types of binary decision variables:

- binary variables describing, whether the empty wagon number l of series nw, which is at the sk station, will be used to complete the order z. In a matrix form:

$$\mathbf{X1} = \begin{bmatrix} x1(l(nw), t_0, sk, z) : x1(l(nw), t_0, sk, z) \in \{0, 1\} \\ nw \in \boldsymbol{NW},\ sk \in \boldsymbol{SN}, z \in \boldsymbol{Z},\ l(nw) \in \boldsymbol{L(nw)},\ t_0 \in \boldsymbol{T} \end{bmatrix}$$

where:

$x1(l(nw), t_0, sk, z) = 1$ - when the empty wagon l of series nw, which is at the sk station, will be used to complete the order z, otherwise $x1(l(nw), t_0, sk, z) = 0$,

- binary variables describing whether the loaded wagon l from series nw, present at sk station, will be used to complete an order number z. In a matrix form:

$$\mathbf{X2} = \begin{bmatrix} x2(l(nw), t_0, sk, z) : x2(l(nw), t_0, sk, z) \in \{0, 1\} \\ nw \in \boldsymbol{NW},\ sk \in \boldsymbol{SN}, z \in \boldsymbol{Z},\ l(nw) \in \boldsymbol{L(nw)},\ t_0 \in \boldsymbol{T} \end{bmatrix}$$

where:

$x2(l(nw), t_0, sk, z) = 1$ - when the loaded wagon l of series nw, which is at the sk station, will be used to complete the order z otherwise $x2(l(nw), t_0, sk, z) = 0$,

- binary variables describing whether the empty wagon l of series nw, which is on the railway line sk, sk' will be used to complete the order z. In a matrix form:

$$\mathbf{Y1} = \begin{bmatrix} y1(l(nw), t_0, n(sk, sk'), z) : y1(l(nw), t_0, n(sk, sk'), z) \in \{0, 1\} \\ nw \in \boldsymbol{NW},\ (sk, sk') \in \boldsymbol{LK},\ z \in \boldsymbol{Z},\ l(nw) \in \boldsymbol{L(nw)},\ t_0 \in \boldsymbol{T} \end{bmatrix}$$

where:

$y1(l(nw), t_0, n(sk, sk'), z) = 1$ - when the empty wagon l of series nw, which is on the railway line sk, sk' will be used to complete the order z otherwise $y1(l(nw), t_0, n(sk, sk'), z) = 0$.

- binary variables describing whether the loaded wagon number l of series nw, which is on the railway line sk, sk' will be used to complete the order z. In a matrix form:

$$\mathbf{Y2} = \begin{bmatrix} y2(l(nw), t_0, n(sk, sk'), z) : y2(l(nw), t_0, n(sk, sk'), z) \in \{0, 1\} \\ nw \in \boldsymbol{NW},\ (sk, sk') \in \boldsymbol{LK},\ z \in \boldsymbol{Z},\ l(nw) \in \boldsymbol{L(nw)},\ t_0 \in \boldsymbol{T} \end{bmatrix}$$

where:

$y2(l(nw), t_0, n(sk, sk'), z) = 1$- when the loaded wagon l of series nw, which is on the railway line sk, sk' will be used to complete the order z otherwise $y2(l(nw), t_0, n(sk, sk'), z) = 0$.

3.4 Constraints and Criterion Function

Model limits (boundary conditions):

- guarantee of execution of submitted and approved customer orders,
- moving wagons for a specific order may start after reserving empty wagons for this order. Empty wagons must arrive before they can be loaded,
- the sum of empty wagons at stations, full wagons at stations, wagons at the stations during loading and unloading, loaded wagons in route and empty wagons in route must not exceed the maximum number of wagons owned by the carrier,
- the number of series wagons, which are used to execute the order, cannot be higher than the total number of series wagons that are in possession of the carrier,
- the number of wagons in order z cannot exceed the maximum number of wagons limited by technical and operational constraints,
- the sum of empty wagons at the station, full wagons at the station, the wagons being loaded, the wagons being unloaded must not exceed the capacity of that station.

For the above mentioned it is necessary to find plans for relocation of loaded and empty wagons which minimize the criterion function (cost minimization - simplified formula (1)), *kdp* - the travel cost of empty wagons to the ordering station, *kdl* - the travel cost of loaded wagons to the forwarding station. The criterion function takes into account two situation: the first part of the pattern (1) in which the wagons (empty or loaded) are located in the stations (decision variables: **X1** for *kdp* and **X2** for *kdl*) and the second part of the pattern (1) in which the wagons (empty or loaded) are located on the route between stations (decision variables **Y1** for *kdp* and **Y2** for *kdl*).

$$
\begin{aligned}
&\forall t_0 \in \boldsymbol{T} \\
\mathbf{F}(\mathbf{X1},\mathbf{X2},\mathbf{Y1},\mathbf{Y2}) &= \sum_{sk \in \boldsymbol{SK}} \sum_{z \in \boldsymbol{Z}} \sum_{sk'' \in \boldsymbol{SK}} (kdp(sk,t_0,sk'',z)\cdot \mathbf{X1} + kdl(sk,t_0,sk'',z)\cdot \mathbf{X2}) \\
&+ \sum_{(sk,sk') \in \boldsymbol{LK}} \sum_{z \in \boldsymbol{Z}} \sum_{sk''' \in \boldsymbol{SN}} (kdp((sk,sk'),t_0,sk''',z)\cdot \mathbf{Y1} + kdl((sk,sk'),t_0,sk''',z)\cdot \mathbf{Y2}) \rightarrow \min
\end{aligned} \tag{1}
$$

4 The Algorithm of the Method

The method of relocating loaded and empty wagons, developed in this dissertation, concerns the creation of a transport plan, i.e. the plan for the transport of loaded and empty wagons on the rail network. It is designed for the railway companies, which carry out transport services.

To fulfill orders submitted by customers, the carrier must allocate wagons for these orders. The algorithm for the relocation of loaded and empty wagons is presented on Fig. 3.

The first stage of the method - entering the data. In order to solve the task it is necessary to prepare the input data, including data on the characteristics of the transport network, availability of resources, data on the demand for transport of goods, as well as

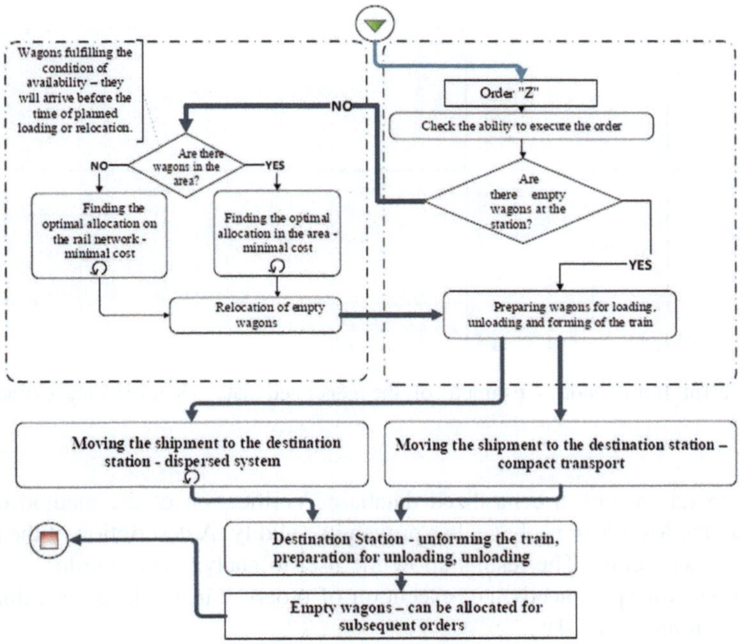

Fig. 3. Algorithm for transporting loaded and empty wagons

boundary conditions related to this process. The second stage is the preparation of plans to move the loaded and empty wagons. Here the task of optimization is to be solved.

The next stage contains the results analysis. At this point, the carrier evaluates if the solution was rational and (most importantly) whether the result provides all orders with wagons. The method is designed to assist the carrier in making decisions.

Hence, the quality of the result and the fulfillment of expectations need to be determined. Organizing the process of moving the loaded and empty wagons is an iterative process and may require the carrier to modify the input data or adjust the calculation method parameters according to the results obtained.

5 Implementation of the Model in the Form of Simulation Package ModPCar - Case Study

The ModPCar program allows the generation of a large number of plans variants. Relocation of full wagons is correlated with the plans for moving empty wagons. The package is dedicated to railroad carriers operating on an extensive area with a large wagon stock and a wide range of transport orders. ModPCar package performs tasks which can be divided into two areas. The first group of functions is designed for creating, saving, printing, and editing data files. The second group is related to the input of data files, performing the simulation and presenting the results. The program was prepared in a demo version. With production implementation, it can be used for

Plan of loaded cars relocation										Plan of empty cars relocation					
Shipment No.	From station	To station	Number of cars	Cars deficit	Series	Scheduled time of arrival of empty cars	Wagon providing	Departure	Arrival	Shipment No.	Departure	Arrival	From station	To station	Number of cars
15	ST_88	ST_61	26	0	Eaos	2010-10-06 16:27	2010-10-06 20:00	2010-10-08 06:04	2010-10-10 23:43	15	2010-10-06 16:27	2010-10-06 16:27	ST_88	ST_88	26
23	ST_89	ST_94	40	0	Eas	2010-10-05 19:16	2010-10-06 08:00	2010-10-07 19:33	2010-10-09 03:08	23	2010-10-04 10:40	2010-10-05 19:16	ST_94	ST_89	17
										23	2010-10-04 12:01	2010-10-05 19:16	ST_88	ST_89	23
25	ST_119	ST_108	29	0	Eas	2010-10-06 22:00	2010-10-07 02:00	2010-10-07 14:00	2010-10-08 08:13	25	2010-10-06 21:18	2010-10-06 22:00	ST_88	ST_119	29

Fig. 4 Excerpt from report - example of the generated data. (Source: data extracted from ModPCar)

network operation with a centralized database. Verification of the method on a real data, using the ModPCar package, has proven its validity. A description of the resulting tables is shown below. The results allow the user to analyze every order.

Satisfying transport needs, i.e. execution of orders, can be done according to the following variants (Fig. 4):

- option 1 - transport wagons are located at the forwarding station and do not need to be brought from other stations,
- option 2 - wagons for freight demand are imported from different stations which are not a sending station,
- option 3 - wagons are imported only from one station and it is not a sending station.

Option 1 - freight wagons are located at the sending station and do not need to be brought from other stations. Example (shipment number 15 - Fig. 4): a customer has requested the transport of 26 Eas cars from ST_88 to ST_61. They ordered the wagons to be provided for siding 2010-10-06 at 20.00. To execute this order, the system reserves the wagons that will be empty at ST_88 at that time. A loaded train with 26 cars from ST_88 station should leave 2010-10-08 at 06.04 and arrive at station ST_61 2010-10-10 at 23.43.

Option 2 - wagons for transport demand are imported from different stations which are not the sending station. Example (parcel number 23 - Fig. 4): a customer has requested to transport 40 wagons of the Eas series in relation ST_89 to ST_94. He ordered the wagons to be provided for siding 2010-10-06 at 08.00. To fulfill the order, 17 cars needed for loading should be departed from station ST_94 2010-10-04 at. 10.40. A train with 17 empty wagons should arrive at station ST_89 2010-10-05 at 19.16. Twenty-three missing wagons are available at ST_88. A loaded train with 40 cars from ST_89 station should leave 2010-10-07 at 19.33 and arrive to ST_94 station 2010-10-09 at 03.08.

Option 3 - wagons for transport demand are only imported from one station which is not a sending station. Example (shipment number 25 - Fig. 4): a customer has requested the transport of 29 wagons of the Eas series in relation ST_119 to ST_108.

He ordered the cars to be provided for the siding 2010-10-07 at 02.00. In order to process the request, 29 empty cars should be relocated from station ST_88 and should leave 2010-10-06 at. 21.18. A train with 29 empty cars should arrive at station ST_119 2010-10-06 at 22.00. After loading, the train should leave 2010-10-07 at 14.00 and arrive at station ST_108 2010-10-08 at 08.13.

6 Conclusion

The ModPCar computer application facilitating the process of decision support in the organization of railway transport may find use, among others, for rail freight carriers.

The plan of moving loaded and empty wagons is prepared based on the transport orders submitted by the buyers of transport services. The main advantage of the proposed approach for planning the movement of empty and empty wagons on the rail network is the ability to test different variants of the plan, allowing one to reconcile, plan, and make adjustments using one system. The quality of the input data is decisive for the correctness of the heuristic algorithm and its practical use.

Obtaining the data required by ModPCar applications is possible in companies that have developed rolling stock tracking systems. As a result, a new tool has been developed to support planning the movement of full wagons together with the planning of relocating empty wagons needed to meet transport orders demand.

References

1. Ambroziak, T., Jacyna, M.: Queueing theory approach to transport process dynamics, Part 1. Dynamics of transport network connections. Arch. Transp. **14**(4), 5–20 (2002)
2. Jacyna, M.: Multicriteria evaluation of traffic flow distribution in a multimodal transport corridor, taking into account logistics base service. Arch. Transp. **10**(1–2), 43–66 (1999)
3. Law Office of the Parliament: Uniform Text of the Act of 28 March 2003 on Rail Transport (Journal of Laws of 2013, Item 1594) with Amendments. Law Office of the Parliament, Warsaw (2003)
4. Midedelkoop, D., Bouwman, M.: Train network simulator for support of network wide planning of infrastructure and timetables. In: Brebbia, C.A., Allan, J., Hill, R.J., Sciutto, G., Sone, S. (eds.) Computer in Railways VII, pp. 267–276. WitPress, Southampton (2000)
5. Cacchiani, V., Caprara, A., Toth, P.: Scheduling extra freight trains on railway networks. Transp. Res. Part B **44**(2), 215–231 (2010)
6. Mu, S., Dessouky, M.: Scheduling freight trains traveling on complex networks. Transp. Res. Part B **45**(7), 1103–1123 (2011)
7. Vansteenwegen, P., Van Oudheusden, D.: Developing railway timetables which guarantee a better service. Eur. J. Oper. Res. **173**(1), 337–350 (2006)
8. Caprara, A., Kroon, L., Monaci, M., Peeters, M., Toth, P.: Passenger railway optimization. In: Handbooks in Operations Research and Management Science, vol. 14, pp. 129–187. Elsevier, Amsterdam (2007)
9. Goossens, J., Van Hoesel, C., Kroon, L.: On solving multi-type railway line planning problems. Eur. J. Oper. Res. **168**, 403–424 (2005)

10. Kuo, A., Miller-Hooks, E., Mahmassani, H.: Freight train scheduling with elastic demand. Transp. Res. Part E **46**(6), 1057–1070 (2010)
11. Mees, A.I.: Railway scheduling by network optimization. Math. Comput. Model. **15**(1), 33–42 (1991)
12. Shi, M., Maged, D.: Scheduling Freight trains traveling on complex networks. Transp. Res. Part B **45**(7), 1103–1123 (2011)
13. Wolfenburg, A.: New version of the BBS method and its usage for determining and scheduling vehicle routes. Arch. Transp. **31**(3), 83–91 (2014)
14. Glover, F.: Future paths for integer programming and links to Artificial intelligence. Comput. Oper. Res. **13**(5), 533–549 (1986)
15. Steuer, R.: An interactive multiple objective linear programming procedure. TIMS Stud. Manag. Sci. **6**, 225–239 (1977)
16. Zeleny, M.: Multiple Criteria Decision Making. McGraw-Hill, New York (1982)
17. Roy, B.: Decision-aid and decision making. Eur. J. Oper. Res. **45**(2–3), 324–331 (1990)
18. Korhonen, P., Laakso, J.: A visual interactive method for solving the multicriteria problem. Eur. J. Oper. Res. **24**(2), 277–287 (1986)
19. Garcia, B.L., Potvin, J.Y., Rousseau, J.M.: A parallel implementation of the tabu search heuristic for vehicle routing problems with time window constraints. Comput. Oper. Res. **21** (9), 1025–1033 (1994)
20. Fügenschuh, A., Homfeld, H., Schülldorf, H.: Single-car routing in rail freight transport. Transp. Sci. **49**(1), 130–148 (2013)
21. Jacyna-Gołda, I., Izdebski, M., Podviezko, A.: Assessment of the efficiency of assignment of vehicles to tasks in supply chains: a case-study of a municipal company. Transport **31**(4), 1–9 (2016)
22. Jacyna-Gołda, I., Lewczuk, K., Szczepański, E., Murawski, J.: Computer aided implementation of logistics processes—selected aspects. In: Mikulski, J. (ed.) Tools of Transport Telematics. CCIS, vol. 531, pp. 67–80. Springer, Switzerland (2015)
23. Jacyna-Gołda, I.: Decision-making model for supporting supply chain efficiency evaluation. Arch. Transp. **33**(1), 17–31 (2015)

Author Index

E. Macioszek and G. Sierpiński (eds.), *Recent Advances in Traffic Engineering for Transport Networks and Systems*, Lecture Notes in Networks and Systems 21,
DOI 10.1007/978-3-319-64084-6

MIX
Papier aus verantwortungsvollen Quellen
Paper from responsible sources
FSC® C105338

If you have any concerns about our products,
you can contact us on
ProductSafety@springernature.com

In case Publisher is established outside the EU,
the EU authorized representative is:
Springer Nature Customer Service Center GmbH
Europaplatz 3, 69115 Heidelberg, Germany

Printed by Libri Plureos GmbH
in Hamburg, Germany